SpringerBriefs in Earth System Sciences

SpringerBriefs in Earth System Sciences present concise summaries of cutting-edge research and practical applications. The series focuses on interdisciplinary research linking the lithosphere, atmosphere, biosphere, cryosphere, and hydrosphere building the system earth. It publishes peer-reviewed monographs under the editorial supervision of an international advisory board with the aim to publish 8 to 12 weeks after acceptance. Featuring compact volumes of 50 to 125 pages (approx. 20,000—70,000 words), the series covers a range of content from professional to academic such as:

- A timely reports of state-of-the art analytical techniques
- bridges between new research results
- snapshots of hot and/or emerging topics
- literature reviews
- in-depth case studies

Briefs are published as part of Springer's eBook collection, with millions of users worldwide. In addition, Briefs are available for individual print and electronic purchase. Briefs are characterized by fast, global electronic dissemination, standard publishing contracts, easy-to-use manuscript preparation and formatting guidelines, and expedited production schedules.

Both solicited and unsolicited manuscripts are considered for publication in this series.

Suchismita Satapathy · Meghana Mishra ·
Manas R. Das

Sustainable Waste Management Practices for the Mining Sector Through Recycling of Mining Waste

Springer

Suchismita Satapathy
School of Mechanical Engineering
KIIT Deemed to be University
Bhubaneswar, Odisha, India

Meghana Mishra
KIIT School of Management
KIIT Deemed to be University
Bhubaneswar, Odisha, India

Manas R. Das
School of Mechanical Engineering
KIIT Deemed to be University
Bhubaneswar, Odisha, India

ISSN 2191-589X ISSN 2191-5903 (electronic)
SpringerBriefs in Earth System Sciences
ISBN 978-3-031-67931-5 ISBN 978-3-031-67932-2 (eBook)
https://doi.org/10.1007/978-3-031-67932-2

This Springer imprint is published by the registered company Springer Nature Switzerland AG
The registered company address is: Gewerbestrasse 11, 6330 Cham, Switzerland

Preface

Waste management, in the mining industry, is a topic of concern among all developing countries. Mining industries are a very important type of industry as they help in the industrialization and progress in terms of new product and process development, in providing employment to countries, and are also as a way of showing scientific and technological advancement. But the unfamiliarity of these industries is its waste production and waste management. The waste produced is tailing waste, waste rock, overburden waste, wastewater, dust and other mining wastes. These wastes have negative impact on earth if remain utilized. The dust, fly ash and poisonous gas generated create air pollution; the overburden, fly ash, tailing waste, red mud, etc., pollute the soil by losing its fertility and making it barren land. The wastewater, used chemicals and garbage of the mining industry pollute water and make the life of living and water bodies in danger. Hence, if the mining wastes are not properly utilized, it creates health and environmental issues. The solutions to these issues are proper waste segregation, use of filtration systems to help in waste management and planting trees and rehabilitation of mining areas. Hence, more research is essential for proper waste management by adopting the recycling process. Recycling of red mud, fly ash, slag, iron ore tailing can be done by producing bricks, concrete, cement, tiles, etc., as building materials for construction industries. Similarly, wastes of mining industry and chemicals may help as required raw material for other industries.

Hence, in Chap. 1, the detailed discussion about challenges faced by mining waste management is discussed. Chapter 2 gives a clear concept on mining waste by literature review. Chapter 3 provides the barriers to making waste management practices, and Chap. 4 provides a design framework to fight with the challenges of waste management. Chapter 5 focuses on application of mining waste as bricks and also tests its noise absorption property. Chapter 6 describes the application of mining waste as ceramic tiles. Chapter 7 clearly explores the addition of mining waste, which enhances the properties of polymer composite. At the end, Chapter 8 concludes, showing the future prospective of mining wastes. It will help the research scholars and researchers to explore new possibility of mining waste utilization.

Bhubaneswar, India

Suchismita Satapathy
Meghana Mishra
Manas R. Das

About This Book

This book is useful for finding mining waste-related problems and challenges of sustainable waste management. It also focuses on the framework or policies against pollution due to mining waste and suggests some ways of resolving these problems by properly utilizing the mining wastes for construction industry.

Contents

About the Authors

Suchismita Satapathy is an associate professor in the School of Mechanical Sciences, KIIT University, Bhubaneswar, India. She has published more than 130 articles in national and international journals and conferences. She has also published many books and e-books for academic and research purposes. Her areas of interest include production operation management, operation research, acoustics, sustainability and supply chain management. She has filled three Indian patents, among them two published. She has more than 15 years of teaching and research experience. She has guided many Ph.D. (three completed, three ongoing), M.Tech. (20) and B.Tech. (49) students. Suchismita has published books such as production operation management (Stadium Press) and MCDM methods for waste management with CRC Press, Innovation, Technology and Knowledge Management with Springer, Soft Computing and Optimization Techniques for Sustainable Agriculture by DEGRUYTER, The PDCA cycle for industrial improvement, applied case studies by Springer, and many of her projects are ongoing.

Scopus Author ID: 55085975400

Meghana Mishra is a research scholar in the School of Commerce in KIIT University, Bhubaneswar, India. She has published more than 6 articles in national and international journals, conferences and book chapters. Her areas of interest include production operation management, economic study of agrisector, sustainability, business process reengineering, ergonomics, etc.

Manas R. Das is a research scholar in the School of Mechanical Sciences at KIIT University, Bhubaneswar, India. He has published more than 4 articles in national and international journals, conferences and book chapters. His area of research is mining waste management.

Abbreviations

AI	Artificial Intelligence
CCTV	Closed Circuit Television
FA	Factor Analysis
HoQ	House of Quality
IoT	Information of Technology
IOT	Iron Ore Tailing
MCDM	Multi-Criteria Decision-Making
QFD	Quality Function DEPLOYMENT
SWARA	Step-wise Weight Assessment Ratio Analysis

List of Figures

List of Tables

Chapter 1
Waste Generation from Mines

1.1 Introduction

In the present scenario, mining operations have increased in a very remarkable scale worldwide due to rapid growth of population and accordingly increased in demand of mineral resources to satisfy its needs. Extraction, beneficiation and processing operations in mines can generate huge amounts of wastes which can cause various adverse effects on environment and can also hinder economic growth of society. Generally, there are two popular mining techniques i.e. opencast and underground. Out of these two, the Opencast mining is more pollution-intensive as it generates 8–10 times more waste compared to the underground mining. Taking for instance, as per than about 3500 operational mining wastes are produced globally. Worldwide, more than about 100 billion tons solid wastes are produced every year from the initial manufacture of goods made of metals and minerals. Similarly, in India the estimated annual solid mining waste is more than about 170 million tons which is expected to exceed substantially to 300 million tons per year with the increase in the production of various minerals. So, taking into consideration the severe environmental and climatic impacts of the gigantic volume of mining waste, its mitigation measure is a serious concern for the mankind.

1.2 Waste of Mines and Its Effect

Wastes are produced at various stages of mining operations, starting from the process of exploration and ending with the closure of mines. Many types of hazardous wastes are generated in a mine ranging from waste rocks, tailings, mine water, processing wastes and chemical wastes.

Waste rocks-heterogeneous materials that moved to obtain the ore mineral in Fig. 1.1a, include overburden (overlying material of the mining area) but these

S. Satapathy et al., *Sustainable Waste Management Practices for the Mining Sector Through Recycling of Mining Waste*, SpringerBriefs in Earth System Sciences,
https://doi.org/10.1007/978-3-031-67932-2_1

exclude topsoil and other soil materials that are reused in reclamation. As waste rocks are exposed to air and water, so a weathering process is carried out which can contribute to generate acid rock drainage or acid mine drainage with high metal contents, especially sulphides, causing threats to the aquatic ecosystem.

Tailings-residual materials generated from the enrichment process in slurry forms with high water contents in Fig. 1.1b.

Mine water-groundwater or precipitation that infiltrates to mines during the extraction process. In mining operations water is often recycled and reused. The water is gathered and eventually escaped to the surrounding environment. In some cases, the water needs to undergo many treatments before escaping back to the surrounding environment.

Processing wastes-residues from processing after beneficiation, such as smelting and electrolytic refining operations.

The mining may occasionally use specific chemicals, such as cyanide, during the phase of processing. Because of their extreme toxicity, even in a very small quantity, that may create a problem.

These mining wastes can cause many negative impacts in the surrounding region and also in societal fabric. for example: land quality degradation, ecological alteration, water quality deterioration, air pollution, ground water shortage, modification in the infrastructure systems, insensible deforestation, demographic migration, economic burden, climate change and heat waves in the urban areas are some of the most prominent issues. In addition, the mining sector generates fossil fuels, the greatest source of carbon emissions in the world, and emits carbon into the atmosphere through energy usage and transportation means. For a long time, span mines still have some of their effects on the surrounding environment even after their complete closure. Often abandoned mines cause acid mine drainage that ultimately results in mixing of heavy metals with the nearby water bodies and consequently impacting the plant, animal and human health in an adverse manner who consume it. Moreover, the environmental effects of mining weaken the potential of women at large to supplement for their families with clean water and nutritious food due to increase in work burden. In order to reconcile economic efficiency with the needs for environmental sustainability, energy requirements, hazards to the environment and human health, demands on water resources and necessary technologies, mining wastes must be carefully handled, managed, and controlled.

1.3 Reuse of Wastes and Benefits

Reuse of mine waste is the method of applying mine waste in its aggregate or in part in a new way without undergoing any kind of initial reprocessing. On the other hand, mine wastes recycling refers to the process of extracting fresh, valuable resource elements or using the waste as feedstock in order to transform whole or part of the mine waste into a useful product after some degree of reprocessing. Waste rock can be recycled and reused as a source of minerals and metals as well as for backfilling,

Fig. 1.1 **a** Waste rocks. **b** Tailings

landscaping, road pavement, embankment, foundation and building construction, cement manufacturing and as soil additive. Mine waste water is primarily used for metal recovery from AMD water, dust reduction, mineral processing, industrial, agricultural and domestic use. Mill tailings are the primary processed wastes generated from mines. Several researches on the usage of mill tailings have verified their use as clay-rich tailings for the production of bricks, cements, floor tiles, sanitary ware, porcelains as well as backfilling materials in underground mines. Similarly iron-rich tailings combined with fly ash and sewage sludge used for making light weight ceramics, phosphate-rich tailings for production of phosphoric acid, manganese-rich tailings used in agro-forestry and as coatings, bauxite-rich tailings as source of alum, used in glass and ceramic industry due to its high alkalinity and also used for treatment of AMD and dairy waste water, copper-rich tailings as paint extenders are some common examples. Since the grades and quality of such ores have declined over the past decades, reprocessing of such ores is becoming very frequent and popular. Reusing waste from mining and mineral processing operations can help reduce the negative environmental impacts relating to disposal; nevertheless, some recycling and reused strategies could actually lead to create new and severe environmental issues. Numerous methods, including ecological risk assessment, life cycle assessment, sustainability operations evaluation, and ecological footprint estimation can be used to evaluate the total environmental cost. However, the most important factor in determining if a certain reuse method is feasible or not is economic cost–benefit analysis.

Ensuring that wastes from mining and mineral processing remain physically, geographically, chemically and radiologically stable should be the major objective when it comes to their disposal. If this is not achievable, the wastes needs to be segregated and kept from interacting with the environment. It is quite challenging to find out a single, effective method to recycle all types of wastes from mining and mineral processing. Every type of waste has a suitable reuse method, These methods can even change depending on the environmental conditions of the surrounding area (e.g. Proximity to drinking water depth of mining activities etc.). when reintroducing mining and mineral processing wastes to the subsurface, precautions must be taken to ensure that the contaminants do not disintegrate in the food chain or water supplies. Incorporating appropriate environmental monitoring and evaluation studies into the reuse design is always recommended. Table 1.1 shows mining wastes their sources, environmental impacts and potential use.

Table 1.1 Showing mining wastes their sources, environmental impacts and potential uses

Mine Wastes	Descriptions	Generated from	Environmental impacts	Potential uses
Waste rock	Rock debris need to be removed to obtain the ores	Underground mines	Surface water, ground water, air and soil	Can be used in tailing dams and backfilling
Tailings	Materials remain after separating the useful and economic ores	Proportion of ground and crushed ore	Surface water, ground water, air and soil	Can be used for construction materials
Mine water	Negatively affected water by mining processes	Infiltrates to the excavated area	Surface water, ground water, soil	Can be recycled and reused as drinking water and dust reduction
Overburden	Surface coating materials of ores	Surface mining	Soil, land	Can be used in dumping and backfilling

Chapter 2
Literature Review on Waste Generation from Mines

2.1 Introduction

During the time of separation of the ores from the ore bodies by blasting process, the waste materials which surround the ores are removed. Similarly, extraction of different minerals from the Earth's crust produces large volumes of wastes. These wastes useless and uneconomical which unnecessarily occupy large mining areas. If, they are improperly disposed of into the surrounding ecosystem, can cause serious threats. So, new strategies must be implemented scientifically for proper relocation, management and disposal of these wastes to incorporate with the existing techniques. At the same time modification and upgradation in the existing waste management techniques are constantly required. Hence, an extensive literature review is essential to discuss proper management of mining wastes with the increase in volume of mineral requirement and waste volume accordingly.

2.2 Literature Review on Waste Generation from Mines

The waste materials generated from different types of mines depend on the ores, topography of miningsites, mining techniques (e.g. Surface mining, underground mining) and the rate of extraction of the ore. The waste materials generally consist of soil, laterites and others. Wastes from mines generally occurs in the form of waste rocks, mining waste water, stylings, sub-grade ores, leached ores and some quantities of toxic chemical and radioactive wastes etc. In surface mining the wastes are removed to recover minerals by using earth moving machineries or manual means. Similarly, in underground mining the wastes are extracted from the excavation sites. The quantity of waste does not include the quantity of top soil layer since it contains some nutrients. Jain et al. [1] have reported mining sectors contribute to a coun-try's economic development, but they produce large volume of solid wastes, including

S. Satapathy et al., *Sustainable Waste Management Practices for the Mining Sector Through Recycling of Mining Waste*, SpringerBriefs in Earth System Sciences,
https://doi.org/10.1007/978-3-031-67932-2_2

overburden and rejects, which can be inadequately disposed of, causing threat to the environment. Shengo [2] have explained that Mining is inevitable for resource-rich countries, providing raw materials and income. However, increasing productivity and modernization of mining activities have led to the extraction of more mineralized rocks. This gives rise to various environmental threats about the management of these wastes, which are made up of waste rocks and mine tailings from ores flotation beneficiation. Jawadand et al. [3] have found every year, billions of tons of solid waste are produced worldwide as byproducts of mining operations. The waste generation varies depending on the mining method, ore type, and processing techniques. Lottermoser et al. [4] Sulfide minerals are common in Earth's crust and can be get in metallic ore deposits, phosphate ores, coal seams, oil shales, and mineral sands. Mining these resources exposes them to oxygenated environments, leading to acid mine drainage. Macías et al. [5] have explained Ac-id mine drainage (AMD) remediation is causing the generation of metal-rich solid residues. Environmental management of these residues requires detailed characterization and stability assessment. Nieva et al. [6] have discovered the sulphide-rich wastes, which contain Fe, Mg, and other minerals, play a crucial role in controlling metal mobility. Additionally, some metals are ad-sorbed onto Fe (hydro)oxides or form less soluble hydroxy sulfates, posing a potential long-term source of heavy metal pollution. Suppes et al. [7] have reported tailings from mine wastes contain valuable minerals, but their potential is often misjudged due to non-standardized assessment methods. Lèbre et al. [8] have intimated metal mining's environmental impact is often dominated by large waste facilities, leading to pollution and mineral loss. The stratification of waste depends on factors like time, extractive strategy, and eco-nomic context. Edwards et al. [9] have explained mining generates significant waste, with an estimated 10 million of tons per year. This waste, contains toxic elements, poses risks to ecosystems and humans. To mitigate these risks, appropriate management or remedial schemes are needed.

Jamieson et al. [10] have discovered mineralogical classification of mine waste can improve risk assessment, guide planning, and optimize remediation. It predicts acidic drainage and metal leaching, and helps understand mineral stability under different redox conditions. Nevskaya et al. [11] have found the Russian Federation's mineral-raw sector is grappling with significant mining waste accumulation due to deterioration of resource quality, geological work conditions, and governance issues. Qureshi et al. [12] have conducted a study on sulphide-bearing waste rock (WRs) from Pakistan's Lakhra coal field and reveals their chemical composition, AMD generating potential, element leachability. The WRs, containing quartz, pyrite, and kaolinite, have mild to strong acid producing potential, threatening natural water quality. Mabroum et al. [13] have found the mining industry generates waste, posing environmental and health hazards. Geopolymerization technology stabilizes polluted waste, reduces greenhouse gas emissions, and offers mechanical, metakaolin substitution, and compressive strength advantages. Miller et al. [14] have explained the challenges of generation and disposal of solid waste in wet and humid environments, focusing on acid mine drainage and waste materials. It proposes a strategy for classifying waste's acid-forming potential and emphasizes the importance of proper sampling. Kuranchie et al. [15] have intimated the potential of mine waste as cost-effective and sustainable construction materials, highlighting the environmental, economic, and land issues caused by its generation. Paktunc [16] have analyzed the environmental challenges in mining industry due to increased metal levels, waste, contaminated groundwater, and soil deposits, primarily caused by tailings, waste rocks, and dusts. Harty et al. [17] have conducted a study on solid waste from ore mining and milling sites found no toxic metal concentrations exceeding EPA criteria, suggesting long-term studies for leachate char-acteristics and closure plans. Šajn et al. [18] have prescribed the chemical composition and quantities of mining and processing waste landfills from historical mining and smelting. It found that around 2.6 gigatons of waste were deposited at 1650 sites, with 55% being conventional mining waste, 37% processing tailings, and 8% metallurgical waste with Serbia and Kosovo having the highest potential. Cobb et al. [19] have found mine wastes and soils mixed with mine wastes contain lead, cadmium, arsenic, and zinc. Akabzaa et al. [20] have conducted the study evaluates acid mine/rock drainage (AMD/ARD) in Ashanti Belt mines using acid base accounting and net acid generation pH tests, confirming two tailing dumps as acid generating.

2.3 Literatures on Hazards Due to Mining Waste

Mining hazards include significant environmental impacts such as waste production, release of toxic substances, air and water pollution, and ecosystem loss due to the emission of hazardous waste (UNDP and UN Environment 2018). Mining causes environmental risks like exposure to harmful particles and contamination and also have some toxicological effects on humans and ecosystems [21]. The hazardous nature of minerals, warn about mining risks and effective remedies for exploration, include bioleaching methods for rare earth elements and copper [22]. Abandoned Pb-Zn mines in The Rio Mannu basin in Italy was identified as the most contaminated, with toxic elements exceeding Italian regulations with Urgent recommended actions include treating adit water, constructing drainage barriers, and consolidating the earthen dam containing the tailing pond to prevent erosion [23]. The Baia Mare region in Romania has mining waste facilities extracting poly-metallic ores, causing environmental risks, including acid rock drainage and water resources [24]. The environmental hazards of sulfide mineral oxidation in mine waste reveal fine-dispersed mill tailings pose the maximum risk, while other sources include coarse particle and ore concentration waste [25]. From Daitari iron-ore mines using satellite data it is revealed that the increase in built-up and mining areas, decreases vegetation land and waste dumps leading to severe environmental impacts [26]. The development of mining industry and intensive use of natural resources in the Murmansk region of Russia leads to a decrease in rich ores reserves, resulting in displaced rock mass, emissions, and toxic elements in water [27]. Environmental hazards in uranium mines on the Michurinske ore field includes man-made sources of natural radiation, such as waste dumps, as sources of ionizing radiation. Waste dumps are considered sources of dust pollution, with uranium accumulation in the dusty fraction [28]. In copper mining wastes in the eastern Amazon pseudo-total concentrations of Cu in all wastes, primarily in artisanal rock waste are highly contaminated by Cu and moderately contaminated by Cr and Ni. Exposure to Cr in the artisanal rock waste is associated with environmental risk [29]. Mining in South Africa is a significant economic contributor but also poses chemical hazards, including dust exposure and toxic substances like platinum, chromium, vanadium, manganese, mercury, cyanide, and diesel particulate [30]. Ecological risk from ionizing contaminants due to mining activity in Nasarawa State, Nigeria, and could be adopted for future work on radioecology of mining [31]. Hazard identification and risk analysis are crucial for any industry to identify and assess associated risks, aiming to reduce them to a tolerable level. Coal mines have high risks than the iron ore mine, and the same was true for medium risk events. This approach helps prioritize addressing the most dangerous situations and minimizing major issue [32]. Waste production and disposal are a growing concern due to their potential hazardous effects on living organisms. So, management strategies are being developed at both government and individual levels. Countries worldwide are working on hazard management to ensure environmental sustainability [33]. Closed mines in the Far Eastern Federal District contains more hazardous tailing dumps with toxic waste accumulation [34].

2.4 Literatures on Reuse of Waste

Mineral production is crucial for economic development, but managing mining wastes is a vital issue. Experimental research has explored the reuse of mining waste rocks as fine aggregates in fly ash-based geopolymers. Physical, chemical, and mechanical tests revealed a homogeneous geopolymer matrix surrounded in the waste rock, governing the system's mechanical behavior. The geocomposites meet the prescribed limit for inert or non-hazardous waste [35]. To ensure a sustainable future, we must prevent, minimize, reuse, and recycle waste. However, sometimes not economically viable, leading to most waste being stored. Significant research is required to develop cost-effective options [36]. Waste material from mining, leading to issues with waste disposal and environmental costs. Sustainable technologies can reduce waste disposal costs and minimize negative impacts on the environment. For instances, abandoned mining in Russia and construction sites in Poland have led to vibrations on neighboring areas. To improve environmental protection and rational use of natural resources, a shift from protective measures to planned management and a mathematical model of geotechnology's impact is needed [37]. The research program REMINE explores the use of Panasqueira mine waste aggregates and fillers in construction materials for transportation infrastructures. The alkali-activated composites from mining/quarrying wastes at, focusing on rheology requirements for various processing techniques. This could lead to less porous and harder aggregates for road paving materials, mitigating the economic and social impact on local communities [38]. Mining and metallurgical processes in South Africa produce significant waste, posing environmental and human health risks. However, these industries can be integrated into a circular economy model, promoting zero waste through recycling and re-use [39]. European mining and quarrying activities generate 55% of total industrial waste, with most being dumped on land or landfill sites. In Portugal, 20% of industrial waste comes from mines and quarries, particularly from Panasqueira mining. The paper discusses the potential for reuse of waste-rock piles from Panasqueira tungsten mines, proposing the development of innovative polymer-based composite materials from non-contaminated waste-rock tailings. These materials can be used for technical-artistic value-added applications, such as conservation, restoration, and architectural interventions [40]. China's underground coal mines produce 95% of total coal production, generating 660 million tonnes of waste annually. 62% is reused and recycled, with low-cost technologies transforming them into environmentally friendly coal briquettes for electricity generation, iron, steel, and cement industries [41]. This study aimed to assess the reuse of mixed mine tailings and waste rock in water-balance covers (WBCs) for geoengineering applications [42]. Masonry units, made of clay or Autoclaved Aerated Concrete (AAC), are widely used in Romanian constructions due to their superior mechanical and thermal characteristics, creating jobs and supporting economic development by integrating mining waste as secondary raw materials in masonry unit production [43]. Minig water can be reused and recycled, including acid mine drainage (AMD), and future recycling practices. Contaminated mine water, including AMD and coal seam gas (CSG), deteriorates land and water ecosystems, causing environmental and public health issues. CSG mining activities alter groundwater quality, posing a threat to agricultural irrigation [44]. China's rapid economic growth has paved way to the production of significant amounts of hazardous mining waste, including coal gangue, sludge, fly-ash, and coal-bed methane. Jincheng Anthracite Mining Group in Shanxi Province has implemented sustainable waste management techniques, reusing and recycling waste in collieries. This has resulted in the conversion of waste into wealth [45]. Coal mine waste rocks can be reused as a sustainable road construction material after investigation of its chemical, mineralogical, geotechnical properties, and acid mine drainage formulation. Similarly, weathered Coal mine waste rocks can be used for embankments and mixed with stabilizers like fly ash and hydraulic road binder. The environmental impact of Coal mine waste rocks is negligible, and the reuse is economically feasible, resulting in lower costs compared to conventional materials [46]. Repurposing of mining waste into concrete aggregate, promoting eco-friendly practices by using recent developments and technological innovations, highlighting the environmental, economic, and practical advantages of this sustainable construction material. So, promoting circular economy principles, reducing environmental impact and advancing resource efficiency [47]. South Africa's waste legislation is obstructing the reuse of mine waste, as it is classified as hazardous and prone to environmental liabilities like acid mine drainage and leaching. The legislation also promotes cradle-to-grave approaches, leaving mine residue trapped between waste and by-product with economic value [48]. Phosphate screening waste rock 0–10 mm (SWR) as alternative aggregates for pavement applications without treatment in capping layer construction for low-volume traffic pavement projects, promoting a transition to a circular economy [49].

References

1. Jain MK, Das A (2017) Impact of mine waste leachates on aquatic environment: a review. Curr Pollut Rep 3:31–37
2. Shengo LM (2021) Review of practices in the managements of mineral wastes: the case of waste rocks and mine tailings. Water Air Soil Pollut 232(7):273
3. Jawadand S, Randive K (2021) A sustainable approach to transforming mining waste into value-added products. Innovations in sustainable mining: balancing environment, ecology and economy. Springer International Publishing, Cham, pp 1–20
4. Lottermoser B, Lottermoser BG (2010) Sulfidic mine wastes. In: Mine wastes: characterization, treatment and environmental impacts, pp 43–117
5. Macías F, Caraballo MA, Nieto JM (2012) Environmental assessment and management of metal-rich wastes generated in acid mine drainage passive remediation systems. J Hazard Mater 229:107–114
6. Nieva NE, Borgnino L, García MG (2018) Long term metal release and acid generation in abandoned mine wastes containing metal-sulphides. Environ Pollut 242:264–276
7. Suppes R, Heuss-Aßbichler S (2021) Resource potential of mine wastes: a conventional and sustainable perspective on a case study tailings mining project. J Clean Prod 297:126446
8. Lèbre É, Corder GD, Golev A (2017) Sustainable practices in the management of mining waste: a focus on the mineral resource. Miner Eng 107:34–42
9. Hudson-Edwards KA, Dold B (2015) Mine waste characterization, management and remediation. Minerals 5(1):82–85
10. Jamieson HE, Walker SR, Parsons MB (2015) Mineralogical characterization of mine waste. Appl Geochem 57:85–105
11. Nevskaya MA, Fedoseev SV, Marinina OA (2016) The analysis of the problems of mining waste products in the mineral resources sector of the Russian Federation. Int J Appl Eng Res 11(16):9018–9025
12. Qureshi A, Maurice C, Öhlander B (2016) Potential of coal mine waste rock for generating acid mine drainage. J Geochem Explor 160:44–54
13. Mabroum S, Moukannaa S, El Machi A, Taha Y, Benzaazoua M, Hakkou R (2020) Mine wastes based geopolymers: a critical review. Cleaner Eng Technol 1:100014
14. Miller SD, Jeffery JJ, Murray GSC (1990) Identification and management of acid generating mine wastes-procedures and practices in South–East Asia and the Pacific regions. Int J Mine Water 9(1):57–67
15. Kuranchie FA, Shukla SK, Habibi D (2013) Mine wastes in Western Australia and their suitability for embankment construction. In: Geo-congress 2013: stability and performance of slopes and embankments III, pp 1443–1452
16. Paktunc AD (1999) Characterization of mine wastes for prediction of acid mine drainage. In: Environmental impacts of mining activities: emphasis on mitigation and remedial measures. Springer, Berlin, Heidelberg, pp 19–40
17. Harty DM, Terlecky PM (1986) Extraction procedure testing of solid wastes generated at selected metal ore mines and mills. Environ Geol Water Sci 8(3):161–171
18. Šajn R, Ristović I, Čeplak B (2022) Mining and metallurgical waste as potential secondary sources of metals—a case study for the West Balkan Region. Minerals 12(5):547
19. Cobb GP, Sands K, Waters M, Wixson BG, Dorward-King E (2000) Accumulation of heavy metals by vegetables grown in mine wastes. Environ Toxicol Chem Int J 19(3):600–607
20. Akabzaa TM, Armah TEK, Baneong-Yakubo BK (2007) Prediction of acid mine drainage generation potential in selected mines in the Ashanti Metallogenic Belt using static geochemical methods. Environ Geol 52:957–964

21. da Silva-Rêgo LL, de Almeida LA, Gasparotto J (2022) Toxicological effects of mining haz-ard elements. Energy Geosc 3(3):255–262
22. Rebello S, Anoopkumar AN, Aneesh EM, Sindhu R, Binod P, Kim SH, Pandey A (2021) Hazardous minerals mining: challenges and solutions. J Hazard Mater 402:123474
23. Adibee N, Osanloo M, Rahmanpour M (2013) Adverse effects of coal mine waste dumps on the environment and their management. Environ Earth Sci 70:1581–1592
24. Modoi OC, Roba C, Török Z, Ozunu A (2014) Environmental risks due to heavy metal pollution of water resulted from mining wastes in NW Romania. Environ Eng Manage J (EEMJ) 13(9)
25. Masloboev VA, Seleznev SG, Makarov DV, Svetlov AV (2014) Assessment of eco-hazard of copper-nickel ore mining and processing waste. J Min Sci 50:559–572
26. De A, Kumar D (2018) Remote sensing and GIS for assessment and prediction of environmental hazards due to mine waste: case study from Daitari iron ore mines, Odisha, India. In: Earth resources and environmental remote sensing/GIS Applications International Society for optics and photonics, p 10790
27. Masloboev VA, Makarov DV, Baklanov AA, Amosov PV, Seleznev, SG (2016) Methods to reduce the environmental hazards of mining and processing of minerals in the Arctic regions. In: Proceedings of the international mineral processing congress
28. Dudar TV, Titarenko OV, Nekos AN, Vysotska OV, Porvan AP (2021) Some aspects of environmental hazard due to uranium mining in Ukraine. J Geol Geogr Geo-ecology 30(1):34–42
29. Yilmaz E (2011) Advances in reducing large volumes of environmentally harmful mine waste rocks and tailings. Gospodarka Surowcami Mineralnymi-Mineral Resour Manage 89–112
30. Utembe W, Faustman EM, Matatiele P, Gulumian M (2015) Hazards identified and the need for health risk assessment in the South African mining industry. Hum Exp Toxicol 34(12):1212–1221
31. Aznar-Sánchez JA, García-Gómez JJ, Velasco-Muñoz JF, Carretero-Gómez A (2018) Mining waste and its sustainable management: advances in worldwide research. Minerals 8(7):284
32. Paithankar A (2011) Hazard identification and risk analysis in mining industry. Doctoral dissertation
33. Qian G, Duanmu C, Ali N, Khan A, Malik S, Yang Y, Bilal M (2021) Hazardous wastes, ad-verse impacts, and management strategies: a way forward to environmental sustainability. Environ Dev Sustain 1–26
34. Krupskaya LT, Golubev DA, Kolobanov KA (2021) Indicators and factors resulting in crisis phenomena related to the environment hazard of accumulated mineral processing wastes at closed mining enterprises. In: IOP conference series: materials science and engineering, vol 1079, no 7. IOP Publishing, p 072011
35. Capasso I, Lirer S, Flora A, Ferone C, Cioffi R, Caputo D, Liguori B (2019) Reuse of mining waste as aggregates in fly ash-based geopolymers. J Clean Prod 220:65–73
36. Lottermoser BG (2011) Recycling, reuse and rehabilitation of mine wastes. Elements 7(6):405–410
37. Ivannikov AL, Kongar-Syuryun C, Rybak J, Tyulyaeva Y (2019) The reuse of mining and construction waste for backfill as one of the sustainable activities. In: IOP conference series: earth and environmental science, vol 362, no 1. IOP Publishing, p 012130
38. Sangiorgi C, Lantieri C, Tataranni P, Castro-Gomes J, Gabriel M (2016) Reuse of mining waste into innovative alkali-activated-based materials for road pavement applications. In: Functional pavement design. CRC Press. pp 1735–1744
39. Matinde E, Simate GS, Ndlovu S (2018) Mining and metallurgical wastes: a review of recycling and re-use practices. J South Afr Inst Min Metall 118(8):825–844
40. Castro-Gomes JP, Silva AP, Cano RP, Suarez JD, Albuquerque A (2012) Potential for re-use of tungsten mining waste-rock in technical-artistic value added products. J Cleaner Prod 25:34–41
41. Anawar HM, Strezov V, Adyel TM (2019) The reuse and recycling of coal mining waste with zero-waste approach by technological development and integrated management for sustainable growth and benefits. Sustain Econ Waste Manage 31–46

42. Gorakhki MH, Bareither CA (2017) Sustainable reuse of mine tailings and waste rock as water-balance covers. Minerals 7(7):128
43. Cobîrzan N, Muntean R, Thalmaier G, Felseghi RA (2022) Recycling of mining waste in the production of masonry units. Materials 15(2):594
44. Anawar HM, Ahmed G, Strezov V, Siddique RF (2019) Recycling, reuse and treatment technologies of mine water for environmental sustainability and economic benefit in mining operations. Sustain Econ Waste Manage 69–84
45. Haibin L, Zhenling L (2010) Recycling utilization patterns of coal mining waste in China. Resour Conserv Recycl 54(12):1331–1340
46. Amrani M, Taha Y, El Haloui Y, Benzaazoua M, Hakkou R (2020) Sustainable reuse of coal mine waste: experimental and economic assessments for embankments and pavement layer applications in Morocco. Minerals 10(10):851
47. Yu H, Zahidi I, Chow MF, Liang D, Madsen DØ (2024) Reimagining resources policy: synergizing mining waste utilization for sustainable construction practices. J Cleaner Prod 142795
48. Haywood LK, De Wet B, de Lange W, Oelofse S (2019) Legislative challenges hindering mine waste being reused and repurposed in South Africa. Extr Ind Soc 6(4):1079–1085
49. Amrani M, Taha Y, Kchikach A, Benzaazoua M, Hakkou R (2022) Field and economic studies on mine waste: sustainable reuse as aggregates for low traffic pavement structure. Sustainability 14(19):12540

Chapter 3
Barriers of Sustainable Mining Waste Management Practices

3.1 Introduction

Waste management is a challenging issue for all types of industries. Particularly, mining industries are unpopular and always in threatening condition due to lack of proper waste management techniques. Mining waste is hazardous and has a negative impact on both living and non-living. When waste is deposited, it reduces the quality of land, making it infertile, which also affects productivity of land and crop productivity reduced drastically. When these mining wastes are drained into water, the water is polluted and water quality also degrades. The effects of hazardous wastes from mines are also a threat to water animals. Due to open pit mining and transportation, the air quality has also degraded, making the air quality poor. The mining industry faces many objections from non-profit organizations, environmental pollution boards and, to gain reputation, many mining industries open their gates and collaborate with academics and research institutes. To conduct research on using these mining wastes for constructive purposes. Although many research was conducted, still more innovative solutions are essential for the betterment of society.

There are many challenges/obstructions faced by mining industries for proper waste management. Table 3.1 discusses the effects of the mining industry on environment/ecosystem. Ecosystem means air, soil and water, including living and non-living.

3.2 Status of Mining Waste

The status of the global mining waste management market is growing from 172.60 to 208.33 billion tons (https://www.fortunebusinessinsights.com).Mineral and fuel market share is the maximum (83.9%). Helson et al. [1] have written that undesirable materials found near mining after mining exploration are called as mining waste.

S. Satapathy et al., *Sustainable Waste Management Practices for the Mining Sector Through Recycling of Mining Waste*, SpringerBriefs in Earth System Sciences,
https://doi.org/10.1007/978-3-031-67932-2_3

Table 3.1 Effect of mining waste management on earth

Sl. No.	Social	Economic	Envirnmental
1	• Exposure to toxic elements • Waterborn diseases	• Clean up and rehabitation • Loss of agriculture land	• Land degradation • Erosion • Habitat destruction
2	• Displacement of communities	• Resource depletion	• Acid mine drainage • Sedimentation • Chemical contamination
3	• Conflicts • Diseases due to polluted air		• Dust and particulate matter • Emission
4			• Carbon footprint • Methen release

These wastes are large in volume. Jones and Boge [2] have discussed that now the mining industries are responding and trying to waste disposal, to save their status, such that no pollution occurs in the future. Ledin and Pedersen [3] have clearly mentioned that the waste is source of acids and may be the main cause of contamination. Aznar-Sanchez et al. [4] have discussed that more and more research are going on the waste utilization of mining areas. Damoah and Herat [5] have explained that sustainable mining waste management techniques are used to help on waste management. Dold [6] has written that mining industries' reputation decreases due to contamination, including SO_2 and CO_2 emissions and the production of of acid mine drainage (AMD), Ledin and Pedersen [3] have explained that environmental contamination occurs due to mining waste. Zeqiri et al. [7] in this article claims to present statistics about mining products, and also their influence on economic growth (i.e. social well-being) and the environment, as well as critical aspects of mineral waste reduction-reuse-recycling (3R). Famiyeh et al. [8] have explained that the mining industry are now concerned about their environmental impact.

3.3 Research Methodology

To study the challenges/ barriers of sustainable waste management in mining industry,an extensive literature review is conducted. Then a group of standardized questionnaires are configured for the survey in mining industry. The Questions are requested to answer by likert scale(1–5) from experts of mining industry, they are Totally acceptable, acceptable, no statement, not acceptable, Totally not acceptable respectively. To assess the waste management operations, four types of mining industry are studied and a survey is carried on.

Table 3.2 shows Questionnairre for survey. Then the answers of the questions collected in numbers and minitab 17 statistical software is used for doing factor analysis. After factor analysis, to find most important challenge of sustainable waste management of mining industry, SWARA (step wise weight assessment ratio) is used. It is a multi-criteria decision making method that is widely used in many research for ranking. Table 3.3 shows expert opinion collected for this study.

Table 3.2 Questionnairre for survey

Sl. No.	Questionnairre	1	2	3	4	5
1	Legacy waste site clean up strategy					
2	Obeying strict regulations of environmental regulations					
3	Permission of local communities for mining establishment					
4	Advanced technology like sensors for the safety of mining tailing dams					
5	Use of renewable energy to avoid negative impact					
6	Traing and education for miners to use advanced extraction methods					
7	Effective utilization of mining waste for sustainable development					
8	Efficient tailing waste management					
9	Disposal of dry tailing waste to avoid pollution					
10	Minimization of waste generation and utilization of waste					
11	Lack of awareness [9–11]					
12	Economic investment or funding is less for mining waste management [12]					
13	Tracing of geological site for extraction to avoid waste [13]]					
14	Emission of toxic gases while recycling [14]					
15	High energy and resource requirements while recycling [12]					
16	Disposing of waste in a manner that reduces the danger of environmental contamination					
17	Restoring the land to a previous condition					
18	Constant monitoring and reporting to ensure that waste is managed					
19	Regularly check land, water and air quality to measure proper waste management					
20	Reuse and use transportation of waste especially from rural areas					
21	Managing mining waste costs especially for low-profit mining					
22	Lack of actual application and performance of various waste management systems					

Table 3.3 Decision-making experts' characteristics

Area	Qualification	No of experts
Mining engineers	Btech/Master	1
Medical sector	MBBS	1
Environmental sector	Master degree	1
Miners	Master degree	1
Academicians	Doctoral	1

3.4 Result and Discussion

To find challenges of waste management in mining sector minitab 16 is used for statistical analysis. Under the five dimensions like (i.e. mining waste management, environmental impact, regulatory framework, technological limitations) 21 items are selected from 22 items having Chrobanch alpha value of more than 0.5 (Nunnaly 1985), items below 0.5 is discarded and not used for further analysis. Table 3.4 shows factor analysis result.

Then SWARA is used to prioritize challenges of mining industry.

Table 3.4 Factor analysis

	Item	F1	F2	F3	F4
Mining waste management	1	0.724			
	8	0.719			
	9	0.664			
	13	0.545			
Environmental impact	2		0.501		
	10		0.813		
	14		0.865		
	19		0.718		
Regulatory framework	3			0.526	
	11			0.548	
	16			0.710	
Technological limitations	4				0.761
	5				0.682
	6				0.625
	7				0.593

3.4.1 Swara Method Is Used to Prioritize Challenges

Steps used for Swara method

As per the importance, the challenges were sorted.

- The average value (s_j) is measured by considering relatve importance of other parameters
- The Coefficients (k_j)are calculated by using the below formula:

$$k_j = \begin{cases} 1 & j = 1, \\ s_j + 1 & j > 1 \end{cases}$$

- **Further recalculation of weight** (q_j) occurs by following below formula:

$$\text{Qj} = \begin{cases} 1, & j = 1 \\ \frac{q_{j-1}}{k_j} & j > 1 \end{cases}$$

- Lastly comparative weights (w_j) is found by using below formula

$$W_j = \frac{q_j}{\sum_{k=1}^{n} qk}$$

where

n number of the parameter's.

Tables 3.5, 3.6, 3.7, 3.8 and 3.9 shows step wise swara analysis for factors and subfactors and Table 3.10 shows final ranking.

After the Swara ranking From Table 3.5, it is found that waste management ranked first followed by environmental impact, regulatory framework and technological limitations. And among sub criteria Permission of local communities for mining establishment ranked first, Advanced technology like sensors for the safety of mining tailing dams. Ranked second and Lack of awareness ranked third.

Table 3.5 Swara ranking for factors

Parameter	Relative Significance of average values (s_j)	Coefficients $(k_j = +1)$	Re-calculated Weights $\left(w_j \frac{x_j - 1}{k_j}\right)$	Weights $\left(q \frac{x_{j-1}}{\sum w_j}\right)$	Rank
M1		1	1	0.275900784	1st
M2	0.12	1.12	0.892857143	0.246339986	2nd
M3	0.16	1.16	0.862068966	0.237845504	4th
M4	0.15	1.15	0.869565217	0.239913726	3rd

Table 3.6 Final weight of parameters on mining waste management (M1)

Parameter	Relative	Coefficients	Re-calculated	$\left(q\frac{x_{j-1}}{\sum w_j}\right)$ Weights
	Significance of average values (s_j)	$(k_j = +1)$	Weights $\left(w_j\frac{x_j-1}{k_j}\right)$	
1		1	1	0.289141991
8	0.1	1.1	0.909090909	0.262856355
9	0.13	1.13	0.804505229	0.232616244
13	0.08	1.08	0.744912249	0.215385411

Table 3.7 Final weight for environmental impact (M2)

Parameter	Relative	Coefficients	Re-calculated	$\left(q\frac{x_{j-1}}{\sum w_j}\right)$ Weights
	Significance of average values (s_j)	$(k_j = +1)$	Weights $\left(w_j\frac{x_j-1}{k_j}\right)$	
2		1	1	0.284381081
10	0.05	1.05	0.952380952	0.270839125
14	0.1	1.1	0.865800866	0.246217386
19	0.24	1.24	0.698226505	0.198562408

Table 3.8 The final weight of regulatory frame work (M3)

Sub parameter	Relative significance of average value	Coefficients	Recalculated weights	Last weights
3		1	1	0.403136288
11	0.28	1.28	0.78125	0.314950225
16	0.43	1.43	0.699300699	0.281913488

Table 3.9 The final weight of technological limitations (M4)

Parameter	Relative	Coefficients	Re-calculated	Weights $\left(q\frac{x_{j-1}}{\sum w_j}\right)$
	Significance of average values	$(k_j = +1)$	Weights $\left(w_j\frac{x_j-1}{k_j}\right)$	
4		1	1	0.290606654
5	0.08	1.08	0.925925926	0.269080235
6	0.1	1.1	0.841750842	0.244618395
7	0.25	1.25	0.673400673	0.195694716

Table 3.10 Final table

Factors/ subfactors	Weight values	Rank	
1	0.289141991	4th	
8	0.262856355	8th	
9	0.232616244	11th	
13	0.215385411	12th	
2	0.284381081	5th	
10	0.270839125	6th	
14	0.246217386	9th	
19	0.198562408	13th	
3	0.403136288	1st	Permission of local communities for mining establishment
11	0.314950225	2nd	Lack of awareness
4	0.290606654	3rd	Advanced technology like sensors for the safety of mining tailing dams
5	0.269080235	7th	
6	0.244618395	10th	
7	0.195694716	14th	

3.5 Conclusion

This chapter highlighted the waste management barriers and it was found that waste management is the major concern for all most all mining industries all over the world. Hence steps must be taken to frame policies for proper waste management, such that environmental and social safety can be achieved.

References

1. Hilson G, Murck B (2000) Sustainable development in the mining industry: clarifying the corporate perspective. Resour Policy 26(4):227–238
2. Jones H, Boger DV (2012) Sustainability and waste management in the resource industries. Ind Eng Chem Res 51(30):10057–10065
3. Ledin M, Pedersen K (1996) The environmental impact of mine wastes—roles of microorganisms and their significance in treatment of mine wastes. Earth Sci Rev 41(1–2):67–108
4. Aznar-Sánchez JA, García-Gómez JJ, Velasco-Muñoz JF, Carretero-Gómez A (2018) Mining waste and its sustainable management: advances in worldwide research. Minerals 8(7):284
5. Damoah E, Herat S (2022) A review of sustainable management of mining waste. Int J Environ Waste Manage 29(3):342–359
6. Dold B (2008) Sustainability in metal mining: from exploration, over processing to mine waste management. Rev Environ Sci bio/Technol 7:275–285
7. Zeqiri K, Shabani M, Konjuhi A, Kutllovci F (2018) Foster of mining waste recycling and 3R principles in mining industry

8. Famiyeh S, Opoku RA, Kwarteng A, Asante-Darko D (2021) Driving forces of sustainability in the mining industry: evidence from a developing country. Resour Policy 70:101910
9. Batista M, Caiado RG, Quelhas OL, Lima GB, Leal Filho W, Yparraguirre IT (2021) A framework for sustainable and integrated municipal solid waste management: Barriers and critical factors to developing countries. J Cleaner Prod 312:127516
10. Pollans LB (2017) Trapped in trash:'Modes of governing'and barriers to transitioning to sustainable waste management. Environ Planning A 49(10):2300–23
11. Ayçin E, Kayapinar Kaya S (2021) Towards the circular economy: Analysis of barriers to implementation of Turkey's zero waste management using the fuzzy DEMATEL method. Waste Manage & Res 39(8):1078–89
12. Bui TD, Tsai FM, Tseng ML, Ali MH (2020) Identifying sustainable solid waste management barriers in practice using the fuzzy Delphi method. Resour Conserv Recycl 154:104625
13. Sivakumar M, Renuka P, Chitra P, Karthikeyan S (2022) IoT incorporated deep learning model combined with SmartBin technology for real-time solid waste management. Comput Intell 1(2):38
14. Satapathy S (2017)An analysis of barriers for plastic recycling in the Indian plastic industry. Benchmarking: An International J 24(2):415–30
15. Nunnally JC (1978). Psychometric theory (2nd ed.). New York: McGraw-Hill
16. Satapathy S (2020) An investigation on sustainable supply chain management challenges for Indian thermal power plants. Int J Logistics Syst Manage 37(2):173–224

Chapter 4
Designing a Framework for Sustainable Mining Waste Management Practices

4.1 Introduction

With increasing population and industrialization, mining activity has also increased proportionally. To fulfill the demand of minerals, coals for industrialization more and more mines were digged and wastes like overburdens tailing wastes fly ash and waste water were deposited on the earth, polluting water, air and land. In one way, mining industries was trying to raise the speed of the growth of rapid industrialization and in the other way, by creating environmental pollution it was trying to destroy the whole earth. So the degrading effect of the Three primary stages of production can result in mineralized waste in a metal mine. Mineral losses to waste rock, tailing, slag, and leached ore are the consequence of inefficiencies brought on by a variety of reasons that arise at each stage. A fifth waste stream is mine water, which is either naturally occurring or added during a manufacturing stage and comes into contact with solid waste. Metals that have dissolved are present in the waste water that results. mining industries was the point of discussion for all researcher communities.

The researchers were trying to find the solution of mining industries'waste management by recycling/reusing the wastes for the development of society or trying to produce less waste. Utilization of wastes of mining industries' for various purposes may help mining industries to get their prior image. Lebre and Corder [1] have proposed a structure to use mining waste, which will reduce waste deposition.

Alloway [2] and Pérez et al. [3] have described that the waste generated from mining may be overburden, tailing waste, slurry, slag etc. To diminish the adverse effects of waste on earth, a particular tailing pond or dam or tip is created prior in mining area and recycling can be done by properly handling these waste [4, 5]. But these waste if kept like this without proper waste management, contaminates the land, water, air (environment and Eco system). Human beings and animals also were having health hazards due to this contamination [6].

S. Satapathy et al., *Sustainable Waste Management Practices for the Mining Sector Through Recycling of Mining Waste*, SpringerBriefs in Earth System Sciences,
https://doi.org/10.1007/978-3-031-67932-2_4

4.1.1 Research Methodology for Mining Workers

To study the challenges of mining waste management and provide a systematic framework for waste management, a standard framework is designed and expert suggestions were collected on scales of 1–5 (1 = agree, 2 = partially agree, 3 = no opinion, 4 = partially disagree and 5 = disagree).The experts selected were mining executives and managers, academic experts and research scholars). Table 4.1 shows questions for experts and a Table 4.2 shows details of experts on the basis of the of experts contacted, their place, qualification, age and work work experience.

From Figs. 4.1 and 4.2, it is found that maximum data is collected from India and Iron mines. All the workers worked in the same mines for 5–10 years or more and they aged between 20 and 45 years.

Table 4.1 Designer questionnaire

Sl. No.	Question	1	2	3	4	5
1	Economic incentives and financing					
2	Stakeholder engagement					
3	Environmental impact mitigation					
4	Technological innovation					
5	Waste minimization and reduction					
6	Training and capacity building					
7	Continuous improvement and innovation					
8	Regulatory compliance and governance					

Table 4.2 Details of experts

Experts	Nos	Experience (years)	Age	Place	Qualification
Miners	17	5–10	25–50	India, Nigeria, Ghana	Degree and Diploma
Academicians	23	10–15	35–58	India, Nigeria, Papua new guinea, Chez	Ph.D.
Research scholars/ scientist	31	3–5	25–35	India, Nepal, Bangaladesh, Nigeria, Ethopia	M.Tech, Ph.D.
Mining managers	22	5–10	35–65	India, Nigeria, Ghana	B.Sc., B.Tech, M.Tech.
NGO s	8	10–15	35–50	India	MA, M.Sc, Ph.D.

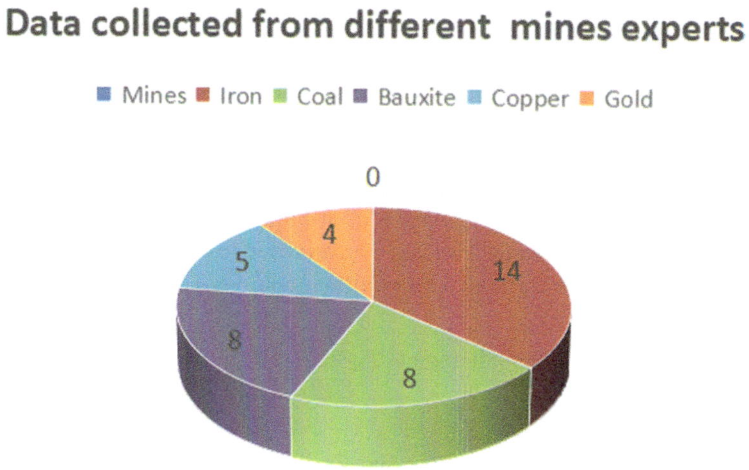

Fig. 4.1 Describes the amount of data collected from different mines of different countries

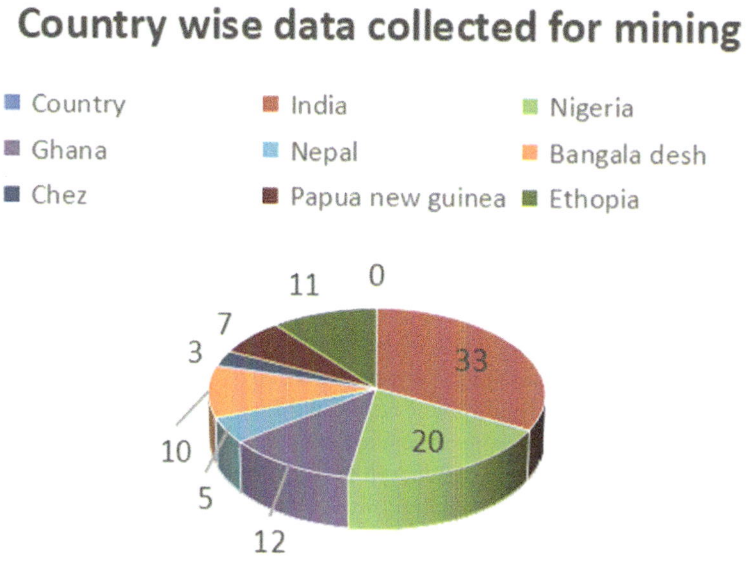

Fig. 4.2 Mines covered for collecting data

4.1.2 Results and Discussion

Quality function deployment is a tool to design a framework which mentions the required policies, design elements and necessary requirements to resolve the raised problems of product or service sector. It designs a HoQ (House of quality) by creating an interrelationship between situational demanding parameters of a system, process, product or service and the suggested solution to resolve the issues. Which is technically called What verses How. In this method, first of all, customer needs is collected and ratings are allotted by customers. The relation between What and How are made by using symbols [O] = 0.8, [ʘ] = 0.6, [Δ] = 0.4 and [●] = 0.2.Which shows better, good, poor and very poor relationships between each other. The steps of QFD to design HoQ framework is explored by Chan and Wu (2005). To find Initial design rating formulas used is $\sum_j^n A_{ij}X_j$, where n = customer needs, A_{ij} and X_j shows relation between customer need and rationg, which is used to design left matrix in HoQ. Similarly, ratings can be calculated by using formula

$$\text{Consumer valuation} = Z_j + \frac{1}{n-1}\left(\sum B_{ij}Z_j\right),$$

where

B_{ij} is called as customers' score and Z_j indicate$_s$ the relationship between customer's requirements.

n is denoted as the number of consumers who participated in this process and put their demands as customer requirements/what. Quality function deployment (QFD) is a tool, which help in the process improvement by helping in improving design of product, process and safety for the betterment of the organization [7]. Memenza et al. [8] has projected QFD application for proper quality management for strawberry farming. Lombardi and Fargnoli [9] have shown the application of the Quality Function Deployment (QFD) techniquein assessing tasks related to hazards to protect the workers from accidents. Yazdani et al. [10] have used QFD in checking customer mannerism in the supplier evaluation process.

Table 4.3 shows customer requirements, and Table 4.4 design requirements as suggested by experts. Table 4.4 shows required HoQ to design a framework for sustainable waste management.

From Table 4.5 Designing house of quality for waste management. It was found that waste minimization and reduction ranked first, environmental impact mitigation ranked 2nd, technological innovation ranked third and regulatory compliance and governance ranked 4th. Waste management and reduction can be done by reducing waste generation, recycling waste and properly segregating waste. Environmental impact mitigation can be done by measuring air, water, soil quality and putting regulations against it, by doing rehabilitation activity in mining areas and conducting awareness problems to discuss environmental impact. Use of advanced technology to reduce waste production, use of AI and automation techniques for proper waste

Table 4.3 Customer requirement (i.e., Barriers of mining waste management)

Sl. No.	Barriers
1	Mining waste management
2	Environmental impact
3	Technological limitations
4	Regulatory framework

Table 4.4 Design Requirement for mining waste management

Sl. No.	Design requirement
1	Economic incentives and financing
2	Stakeholder engagement
3	Environmental impact mitigation
4	Technological innovation
5	Waste minimization and reduction
6	Training and capacity building
7	Continuous improvement and innovation
8	Regulatory compliance and governance

management and use of proper waste treatment methods will improve technological innovation for proper waste management.With regulated policies,audit, stake holders' engagement, proper training for employees, financial /funding support and, lastly, continuous improvement can mitigate all the barriers of waste management (Fig. 4.3) shows HoQ.

4.1.3 Conclusion

Sustainable waste management practices in mining industry can improve the performance of the mining industry, can bring the reputation of the mining industry and can reduce or recycle waste for better use. The generated waste creates air, water and soil pollution. But the importance of mining industries is more for the development of the country, for industrialization. Hence, the focus must be given on reduction of waste production and utilization of waste/ proper management of waste.

Table 4.5 Designing house of quality for waste management

	Requirement	1	2	3	4	5	6	7	8	Customer rating
1	1	▲	▲	▲	○	○	●	▲	▲	6.873
2	4	▲	●	●	○	○	●	●	▲	10.045
3	2	●	●	▲	●	●	●	▲	▲	5.154
4	3	●	▲	▲	▲	○	●	●	▲	7.554
Initial rating		17.1	24.6	29.3	31.2	35.6	34.24	28.93	43.3	
Revised Rating		28.93	32.3	41.1	39.1	43	38	28.63	27.65	
Normalized Rating		0.672790698	0.751162791	0.955813953	0.909302326	1	0.88372093	0.665813953	0.643023256	
		6th	5th	2nd	3rd	Ist	4th	7th	4th	

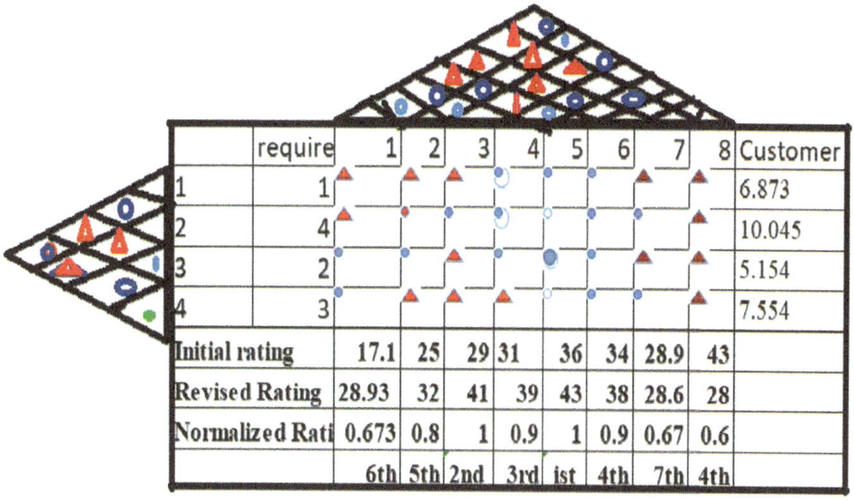

Fig. 4.3 HoQ for waste management

References

1. Lèbre É, Corder G (2015) Integrating industrial ecology thinking into the management of mining waste. Resources 4(4):765–786
2. Alloway BJ (1995) Heavy metals in soils. Blackie, Glasgow, UK. ISBN 0751401986
3. Pérez Cebada JD (2016) Mining corporations and air pollution science before the age of ecology. Ecol Econ 123:77–83
4. Pasariello B, Giuliano V, Quaresima S, Barbaro M, Caroli S, Forte G, Carelli G, Iavicoli I (2002) Evaluation of the environmental contamination at abandoned mining site. Microchem J 73:245–250
5. Hudson-Edwards KA, Dold B (2015) Mine waste characterization, management and remediation. Minerals 5:82–85
6. Iribar V, Izco F, Tames P, Antigüedad I, da Silva A (2000) Water contamination and remedial measures at the Troya abandoned Pb-Zn mine (The Basque Country, Northern Spain). Environ Geol 39:800–806
7. Song C, Wang JQ, Li JB (2014) New framework for quality function deployment using linguistic Z-numbers. Mathematics 8(2):224
8. Memenza A, Cauty N, Sotelo F, Ramos E (2020) Standardize strawberry crops by applying the quality function deployment in huaura, Perú. No. 3679. Easy Chair
9. Lombardi M, Fargnoli M (2018) Prioritization of hazards by means of a QFD-based procedure. Saf Secur Stud 163
10. Yazdani M, Hashemkhani Zolfani S, Zavadskas EK (2016) New integration of MCDM methods and QFD in the selection of green suppliers. J Bus Econ Manag 17(6):1097–1113

Chapter 5
Catagorizing Waste and Utilizing Them by Fabricating and Testing Bricks for Various Properties and Solving Noise Issues

5.1 Introduction

Mining is the procedure of removal of earthly minerals from the earth. After extraction of mining waste, many earthly materials are removed, called as overburden, tailing waste,dust particles, waste mining water, slag and heap leach residue. These are the common mining waste, extracted from the earth. Tailings waste is the unwanted and remaining waste from the ore after the mineral has been removed. [1] have discussed the difference between "sand" and "slime" and they vary by particle size. [2] explained that, with wasted ore and pulp, cyanide and sulphates like hazardous materials and gas are also removed after ore extraction. Like other wastes, mining water is also extracted from mines and it contaminates water-bodies. During mining of overburden rock using surface mining techniques, extra unwanted waste is also removed. As per UNEP (United Nations Environment Programme) hazardous waste as waste is solid, sludge, liquid, or gas and that shows chemical reaction, toxicity, explosiveness etc. [3] explained that hazardous materials have negative effects on human health and the environment. [4] have explained the barriers to sustainable waste management in the mining sector. [5] have investigate waste water treatment and show future directions of its utility. [6] have explained about ecologically friendly materials and their promising futures for the next-generation. Many researchers and scientists have done a lot of investigation on the applications of mining waste, among them applications as polymer composites, bricks, ceramic tiles, concrete etc. as building/construction materials. Still, many mining waste applications remain undiscovered. Hence, it is necessary to categorise the types of waste and search for its applicability. During exploration of the utility of mining waste, its various properties were studied, like water absorption, electricity resistivity and noise absorption property. The bricks utilized for construction purposes are not suitable for controlling noise, so different mining wastes are used to fabricate bricks. Overburden and tailing waste are remain unutilized and it decreases the fertility of land and also creating

S. Satapathy et al., *Sustainable Waste Management Practices for the Mining Sector Through Recycling of Mining Waste*, SpringerBriefs in Earth System Sciences,
https://doi.org/10.1007/978-3-031-67932-2_5

environmental pollution. As mining waste is toxic for all living beings. So, proper utilization of mining waste is essential for the development of society.

5.2 Methodology

In this chapter, first of all, a study is carried out to investigate the types of waste released from the mining industry in Fig. 5.1, then the categorization of waste is shown in Table 5.1 for the iron mining industry in accordance with United Nations Environment Programme (UNEP) criteria. To find the hazardous and non hazardous waste, extensive literature review is conducted and iron mines visited. As per the suggestion of experts of mining industry, Table 5.1 shows types of waste based on hazardous and non-hazardous criteria.

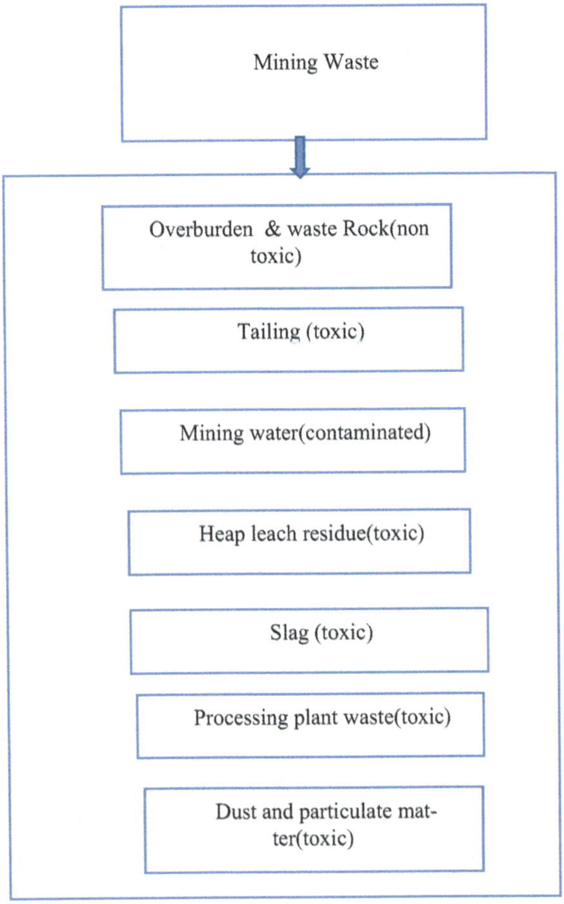

Fig. 5.1 Catagorization of mining waste

Table 5.1 Iron mine waste category [7]

Serial. No.	Column A (Hazardous waste)	Column B (Non hazadrous waste)
Dry tailing	√	×
Wet tailing	√	×
Worthless tailing rock and soil	√	×
Unrequired oils after use	√	×
Use lubricating oil	√	×
Treated charcoal	√	×
Treated (Al_2O_3)	×	√
Molecular size	×	√
Aged balls and rods	×	√
Steel textual matter	×	√
Elastic liners made up of rubber	×	√
Used pumps after abrasion	×	√
Fun used textiles	×	√
Ligament	×	√
Cotton ligaments	×	√
Unused plastics	×	√
Weightlifter bar	×	√
Fibre and plastic items	×	√
Packing materials and boxes	×	√
Material of steel	×	√
Great bar	×	√

The mining waste bricks are made-up without burning of solid fuel, it reduces the release of CO_2 gas [8]. [9] have suggested for manufacturing bricks with the tailing waste of iron mines. The bricks are made with tailing waste of iron ore and with fly ash mixture. At first all the percent of waste like fly ash (55%), lime (15%), gypsum (5%) and sand (10%) are mingled in a wanted ratio for 4–5 min and then the mould is kept 9 × 14 × 29 cm dimension and given pressure to make bricks. These materials undergo a curing process for 14 days and then air dried for 7 days. The iron ore tailing bricks are fabricated by mixing 40% dry tailing with dry clay, blender and water (Behera et al. 2019). A similar procedure is used to make bricks. A 50 mm diameter cylindrical piece is fabricated for sound absorption test. Figure 5.2 shows raw materials for brick preparation. Figure 5.3 shows mixing of raw materials. Figures 5.4 and 5.5 shows fly ash and iron ore tailing bricks after they moulded. Figure 5.6 shows cylindrical mould for sound testing.

Fig. 5.2 Raw materials used for brick preparation

Fig. 5.3 Mixing of raw materials

Fig. 5.4 Iron bricks moulding

Fig. 5.5 Fly ash bricks moulding

Fig. 5.6 Cylindrical moulds for sound testing

5.3 Results and Discussion

For manufacturing of bricks, 950 temperature factor is kept constant for all days (i.e.,7, 14 and 28). For testing these manufactured bricks properties like specific gravity, compressive strength, density, water absorption percentage and porosity are measured for all types of bricks like 7, 14 and 28 days. Table 5.2 shows that the compressive strength of Iron tailing, fly ash bricks are showing maximum compared to others. From Tables 5.2, 5.3 and 5.4, it was clear that, Iron tailing, fly ash brick shows both water absorption percentage and porosity percentage good.

Sound absorption properties depends on the frequencies, composition, thickness, density, porosity etc. Hence, a porous material absorbs more. Noise, as cavities allow sound waves to enter through them. If the thickness of the material is more, it can absorb more sound. Mathematically, it can be expressed as:

$$\alpha = \frac{I_{ab}}{I_i}$$

Table 5.2 Bricks after 7 days

Bricks	Temperature (degree centigrade)	Specific gravity	Compressive strength in mpa	Bluk density (gm/cm^3)	Porosity percentage	Water absorption percentage
Iron tailing, fly ash	950	1.88	29.70	1.97	25.13	11.10
Fly ash	950	1.68	12.09	1.41	16.31	5.91
Iron tailing (40%) (Behera et al.)	950	1.26	24.60	1.82	17.13	14.196

Table 5.3 Bricks after 14 days

Bricks	Temperature (degree centigrade)	Specific gravity	Compressive strength in mpa	Bluk density (gm/cm^3)	Porosity percentage	Water absorption percentage
Iron tailing, fly ash	950	1.47	29.70	1.89	34.13	17.14
Fly ash	950	1.22	15.89	1.22	14.47	8.02
Iron tailing (40%) (Behera et al.)	950	1.19	25.40	1.82	15.57	14.196

Table 5.4 Bricks after 28 days

Bricks	Temperature (degree centigrade)	Specific gravity	Compressive strength in (mpa)	Bluk density (gm/cm^3)	Porosity (%)	Water absorption (%)
Iron tailing, fly ash	950	1.71	24.70	1.57	35.99	14.234
Fly ash	950	1.29	23.44	1.27	18.45	9.86
Iron tailing (40%) (Behera et. al.)	950	1.11 lePara>	25.40	1.82	17.15	14.196

where, I *ab* is sound absorption occurs when, and I_i is intensity of incident sound. If all the sounds are absorbed by the material, α value will remain 1. Hence, it is considered that an alpha value is more than 0.75 are good sound absorbing material.

Fig. 5.7 Sound measuring set up

Table 5.5 Noise testing

	500	1000	1500	2000	2500	3000	3500	4000	5000
Fly ash	0.5	0.48	0.45	0.46	0.48	0.5	0.57	0.63	0.71
IoT	0.67	0.59	0.56	0.6	0.67	0.7	0.72	0.74	0.78
Fly ash + IOT	0.64	0.85	0.91	0.9	0.84	0.69	0.79	0.75	0.81

To measure the sound absorption coefficient of a material, impudence tube method is used. A glass tube with both ends capped is used for measuring sound. A speaker is inserted at one end and material is kept inside without any gap and testing is done at a frequency range of 500–5000 Hz. Figure 5.7 shows set up for sound testing.

From Table 5.5 it is found that bricks, which are fabricated with iron ore tailing & fly ash show the best results. Then the water absorption test was conducted by putting iron ore tailing & fly ash cylindrical piece for 48 h inside water. Then further noise testing was conducted

Then electricity resistivity test was conducted, as per Fig. 5.8 and fly ash and iron ore tailing bricks are bad conductors of electricity.

Fig. 5.8 Electricity restive test

5.4 Conclusion

For proper management of iron ore tailing waste and fly ash, the bricks manufactured show all the properties, and all the properties show good result. It is a method to protect the earth by utilizing mining waste and also helping people from noise pollution. In the future, some more experiments could be conducted by changing the shape of the brick or designing hollow bricks to reduce noise pollution.

References

1. Eriksson N, Staff BE, Adamek P (2000) The tailings pond failure at the Aznalcóllar mine, Spain1. Environ. Issues Manag. Waste Energy Miner. Prod. :109–16
2. Khanna T (2000) Business groups and social welfare in emerging markets: Existing evidence and unanswered questions. Eur Econ Rev 1:44(4–6):748–61
3. LaGrega MD (2001) Phillip L Buckingham & Jeffrey C E-vans
4. Kalisz S, Kibort K, Mioduska J, Lieder M, Małachowska A (2022) Waste management in the mining industry of metals ores, coal, oil and natural gas-A review. J Environ Manag 15:304:114239
5. Igwegbe CA, Obi CC, Ohale PE, Ahmadi S, Onukwuli OD, Nwabanne JT, Białowiec A (2023) Modelling and optimisation of electrocoagulation/flocculation recovery of effluent

from land-based aquaculture by artificial intelligence (AI) approaches. Environ Sci Pollut Res 30(27):70897–917

6. Choudhary S, Panda T, Behl A (2023) Modelling the linkage between fossil fuel usage and organizational sustainability. J Cleaner Prod 10:413:137440
7. Manas M, Sharma S, Reddy KS, Srivastava A (2023) A critical review on techno-economic analysis of hybrid renewable energy resources-based microgrids. J Eng Appl Sci 70(1):148
8. Shrikant RL et al (2016) Utilization of mine waste in the construction industry—a critical review. Int J Earth Sci Eng 9
9. Tang Y, Yin M, Yang W, Li H, Zhong Y, Mo L, Liang Y, Ma X, Sun X (2019) Emerging pollutants in water environment: Occurrence, monitoring, fate, and risk assessment. Water Environ Res 91(10):984–91

Chapter 6
Use of Mining Waste as Ceramics and Testing for Their Properties

6.1 Introduction

Ceramic tiles are a versatile and popular choice for different construction works. They are made from natural clay and other raw materials. Ceramics properties include hard and strong but brittle, they also endure high temperatures (1000–1600 °C)

The manufacturing of ceramic tiles from mining waste is a new and ecologically responsible approach to the mining and building industries. With a growing global awareness of the need to decrease waste and lower the environmental imprint of various industrial processes, the use of mining waste in ceramic tile production marks an important step towards sustainability and resource conservation.

Fly ash and glass are abundant byproducts of industrial processes, particularly in the mining and energy sectors. The utilization of fly ash and glass in ceramic tile manufacturing offers several environmental and economic benefits. Moreover, ceramic tiles produced from fly ash and glass exhibit favorable properties in terms of durability, aesthetics, and performance, making them suitable for a wide range of applications in the construction industry. With proper formulation and processing techniques, these tiles can meet or exceed industry standards for strength, water absorption, and chemical resistance, while offering unique design possibilities and aesthetic appeal.

In this investigation of ceramic tile manufacture using mining waste, we will look at the essential procedures, materials, and long-term benefits of this novel strategy. We'll also look at the broader ramifications for the mining and construction industries, as well as how this practise fits into the rising worldwide commitment to a greener, more sustainable future.

Kummoonin et al. [1] reviews the current research focusing on the recycling of zinc hydrometallurgy red mud waste (RM), furnace slag (FS), and recycled cullet (RC) for the production of vitrified floor tile products. Incorporation was attempted with the goal of developing novel formulations that are less expensive, have less water absorption, and have a high flexural strength. The results showed that all of

S. Satapathy et al., *Sustainable Waste Management Practices for the Mining Sector Through Recycling of Mining Waste*, SpringerBriefs in Earth System Sciences,
https://doi.org/10.1007/978-3-031-67932-2_6

the industrial wastes (RM, FS, and RC) could be used to make floor tiles. The new goods, which were made up of 70RM, 15FS, and 15RC and burned at 1050 °C, met the requirements of Thailand's industrial standard.

Kim and Park [2] have looked into the cutting-edge synthetic technology that creates wear-resistant glass ceramics from mining waste and smelting leftovers. Suvorova et al. [3] have explored that the mining waste can be used in building industry. Haiying et al. [4] shows that Fly ash is used in this job as a blending agent to make ceramic tile. XRD, IR, and SEM analyses were also used to evaluate the product's macro-performance, microstructure, leaching toxicity, and practical application. Sequential chemical extraction and leaching toxicity analyses were also conducted. It is found that the glazed tile demonstrates exceptional resistance against leaching of heavy metals in accordance with the HVEP standard. These results show that heavy metals are difficult to remove because they are deeply buried in the solid lattice of the product.

Kockal [5] claims that the main challenge is figuring out the process input parameters that will produce a ceramic tile that has been sintered and has little shrinkage, good flexural strength, and low water absorption. In general, a variety of variables, either singly or in combination, can affect the effectiveness of the sintering process, such as temperature, constituent type, and content. As a result, response surface methodology was used to improve the responses of these three input variables (RSM). Analysis of variance (ANOVA) was used to present the results of the test of the factors' and their interactions' relevance for the replies. The fly ash (FA) concentrations and temperature have the strongest relationships with the output parameters.

Bonczyk and Rubin [6] explains the findings of a study on the physical and radiological properties of building ceramics derived from mining waste. However, some waste qualities may preclude such utilization. The qualities of bricks constructed from mining waste—shale obtained from dumps—were investigated in the scope of this investigation. It has been demonstrated that the qualities of these bricks match the criteria set out in numerous standards and legislative acts and may be utilized in construction.

Almeida et al. [7] states that to make ceramic tiles, new eco-friendly mass formulations based on scheelite and kaolin wastes were created. The chemical composition, principal mineralogical phases, and particle size distribution of the starting raw materials (scheelite residue, kaolin residue, feldspar, and plastic clay) were all determined. Linear shrinkage, water absorption, apparent density, apparent porosity, flexural strength, and mineralogical phase identification were used to assess the potential of the mass formulations investigated. The findings were compared to experimental data from the literature and International Technical Standards. The samples studied were found to have adequate characteristics for use as ceramic and porcelain tiles.

Vieira and Monteiro [8] claims to have updated a study of the ideas and technological achievements in the incorporation of industrial solid wastes into red ceramics. The initial waste categories were modified to allow for the examination of a broader range of wastes. A property modifying wastes category was added to replace the previously intended fly-ash and plasticity reducing/plastifying wastes categories, in addition to fuel and fluxing wastes. To be more comprehensive, this new analysis examined industrially important subcategories within each category. As a result, greasy leftovers, blast furnace sludge, and paper mill dregs are among the fuel wastes. Fluxing byproducts include decorative rock rejects, glassy byproducts, and fluxing ashes.

Amrani et al. [9] examines the possible use of coal mine waste rocks for road construction projects. Ogunro et al. [10] shows that research has been done on the use of leftover glass in the production of ceramic tiles in order to reduce manufacturing energy consumption. The foundational element for this inquiry was clay. Tests for cold crushing strength (CCS), water absorption, burnt linear shrinkage, density, and apparent porosity were conducted on the created square tiles. This study showed that, in addition to being utilized in clay products, leftover glass could also be economically employed to make ceramic tiles by decreasing the firing temperature. Because the clay can reduce the firing temperature by 55 °C (about 3.9%), soda lime glass is an affordable sintering aid when used in the production of ceramic wall tiles.

6.2 Research Methodology

Creating ceramics from mining waste is an innovative and sustainable approach that can help manage industrial by-products while producing valuable materials. Several types of mining waste can be used for ceramic preparation, each contributing different properties to the final product. Olgun et al. [11] have suggested that ceramic tiles made up of fly ash have good firing strength, firing shrinkage, water absorption. Vasic et al. [12] have suggested granite waste as a best material, having good water absorption property and cheaper compared to feldspar. Lachibi et al. (2023) suggested that glass waste tiles show good mechanical property. Luo et al. [13] explored that the ceramic tiles prepared from fly ash, glass waste and granite waste shows good sintering property and low firing temperature. Hence proper utilization of mining waste occurs, hence can be used for outdoor tiles.

To fabricate ceramics with the help of mining waste, the properties of ceramic materials were studied (Table 6.1). The materials investigated for the production of ceramic tiles encompassed a diverse range, among which notable components included fly ash, granite waste and glass. These materials were subject to thorough analysis and assessment to evaluate their suitability for incorporation into ceramic tile formulations.

In this work, a manufacturing technique dry press forming method was used to test fly ash additions ranging from 30 to 50 wt%. Tiles were created in square tiles (70 × 70 × 260) mm were utilized for the water absorption and abrasion testing. Here are the steps involved in making ceramic tiles: (i) Mixing: Fly ash and clay were combined in a mixing bowl after being weighed according to the specified amounts. (ii) Moulding: Transfer the powder into the pressing die in an even layer. The maximum pressure is maintained for 30 s prior to release. (iii) Drying: The unfired samples are dried in air for 3 days. The unfired tiles must be as dry as possible at this point in order for

Table 6.1 Waste mining materials for ceramic tiles Kummoonin et al. [1]

	Average particle size (μm) NB	Density (gm/cm^3) B	Temp (°C) B	Flexural strength (MPa) B	Water absorbtion (%) B
Red mud	16.44	1.75	900	20	3
Furnace slag (FS)	29.4	2.88	950	22	4
Recycled cullet (RC)	30.74	2.64	1000	25	4.5
Kaolin	51	1.88	1250	7.9	65
Granite waste	1.5	1.6	1150	35	38
Fly ash	21.2	2.9	1350	61.4	11.48
Sewege sludge	1.2	1.4	40	4.84	93
Glass ceramics	4.8	2.56	700	70	2.7
Limestone	8.4	2.71	1100	10.3	2.7

them to feel warm to the touch. (iv) Firing: After positioning each tile on the kiln shelves (shown in figure), the tiles are burned in a furnace high temperature for 2 h at a maximum temperature of 750 °C. The furnace cools naturally after it has fired and stops heating. Figure 6.1 shows the mixing process, Fig. 6.2 Moulding process and Fig. 6.3 drying process of ceramic tiles.

Fig. 6.1 Mixing of materials

Fig. 6.2 Moulding, process

Fig. 6.3 Dried samples from the mould

6.3 Results and Discussion

In the production of ceramic tiles derived from fly ash and glass, the firing temperature plays a critical role in determining the final properties and characteristics of the tiles. It is noted that a firing temperature of 750 C for a duration of 2 h was employed in the firing process of the ceramic tiles under investigation. This temperature and duration were carefully selected based on preliminary experimentation and consideration of the materials' compositions and desired outcomes. Figure 6.4 shows the furnace used for heating, Fig. 6.5 the samples after being removed from the furnace and Fig. 6.6 shows the samples after cooling naturally.

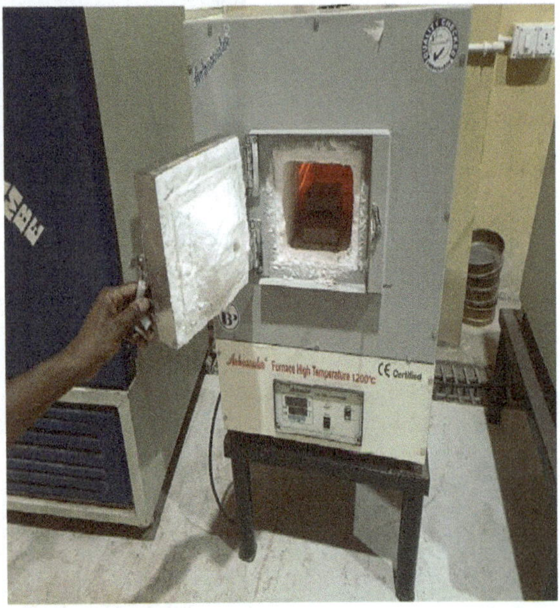

Fig. 6.4 Furnace used for firing

Fig. 6.5 Sample after being removed from the furnace

Fig. 6.6 Sample after it cools

In this study, water absorption analysis was conducted by immersing the ceramic tiles in water for a duration of two hours, followed by allowing them to air-dry overnight. This methodology was chosen to simulate real-world conditions and replicate potential exposure to moisture over an extended period. By subjecting the tiles to a soaking period, any pores or voids within the material are given ample time to absorb water, allowing for accurate measurement of the til's water absorption capacity. Subsequent overnight drying ensures that excess moisture is expelled from the tile's surface, enabling the determination of its residual water content

$$E_v = \frac{m_2 - m_1}{m_1} \times 100$$

M_2 mass after being submerged in water and dried = 160.5 g
M_1 mass being being immersed in water = 157 g
E_v water absorption percentage

In this test the water absorption was found to be 12.23% indicate it as semi-vitreous or vitreous tiles. Then an abrasion test was conducted. At a low firing temperature of 1050 °C. Increasing the firing temperature the water absorption decreases. Where the temperature are given 1100 and 1120 °C the water absorption is reduced to 11.79% and 10.97%, respectively (Fig. 6.7).

The abrasion test indicates the material's resistance to wear and tear under abrasive conditions. If the tile exhibits low weight loss and minimal surface damage after undergoing the abrasion test. The tiles sample underwent an abrasion test and the results are as shown on Table 6.2. According to the results of samples, it suggests that the tile has significant wear of almost 2 mm and material removal during the test

Fig. 6.7 Tile abrasion machine

suggesting the tile would have a not so good abrasion resistance. Figure 6.8 shows Abrasion resistance test results.

The tests conducted on a ceramic tile made from fly ash yielded significant results. The development of ceramic products from fly ash with 25–30% content showcased high strength ceramic composite boards and acid and abrasion-resistant tiles, indicating the versatility and durability of ceramic products incorporating fly ash. These results highlight the potential of fly ash to enhance the properties of ceramic tiles, making them suitable for various applications with improved strength and resistance to wear and abrasion.

Based on these results, it appears that the ceramic tile made from fly ash may have limited abrasion resistance, as evidenced by the substantial weight loss and change in thickness observed during the test. Further investigation and potential modifications to the tile's composition or manufacturing process may be necessary to improve its abrasion resistance and overall durability for practical applications.

Water absorption is an important property of ceramic tiles, as it affects their suitability for various applications, particularly in wet environments. According to the test results the sample A showed us it had a small water absorption percentage

Table 6.2 Showing thickness reduction of sides during abrasion process

	Weight (grams)	1 side (mm)	2 side	3 side	4 side
0 (before abrasion test)	168	25.5	26	26	26.5
1	162	25	25.5	25	25
2	160	24	24.5	24.5	24.5
3	158.5	23.75	24.25	24	24
4	157	23.5	24	23.75	24

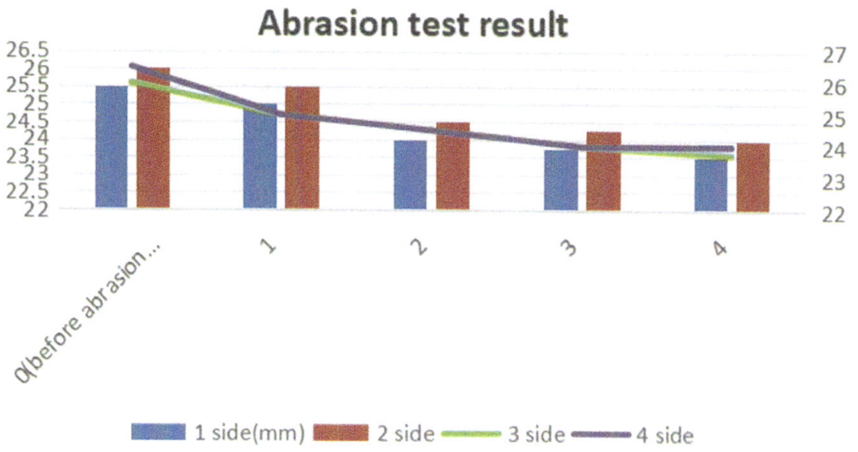

Fig. 6.8 Abrasion resistance test results

that could classify it as a semi vitreous or vitreous tile based on the ASTM C373 standards.

Ceramic tiles made from fly ash and glass ceramic can be used in various applications. The tiles' low water absorption rate and high durability make them suitable for high-traffic areas and wet environments.

6.4 Conclusion

The ceramic tile made from fly ash and glass ceramic has the potential to offer a sustainable and high-performance alternative to traditional ceramic tiles. However, there were some failures in the testing process and the brittleness of the tile after production. These issues may affect the tile's performance and durability, and further research and development are needed to address these challenges.

The ceramic tile industry is facing a growing demand for sustainable and high-performance building materials, and the use of innovative materials, such as fly ash and glass ceramic, can provide a competitive advantage for manufacturers. However, it is essential to ensure that the tiles meet the required standards and performance criteria.

References

1. Kummoonin N, Jaimasith M, Thiemsorn W (2014) Fabrication of ceramic tiles from industrial wastes. Suranaree J Sci Technol 21(2):65–77

2. Kim Y, Park H (2020) A value-added synthetic process utilizing mining wastes and industrial byproducts for wear-resistant glass ceramics. ACS Sustain Chem Eng 8(5):2196–2204
3. Suvorova OV, Selivanova EA, Mikhailova JA, Masloboev VA, Makarov DV (2020) Ceramic products from mining and metallurgical waste. Appl Sci 10(10):3515
4. Haiying Z, Youcai Z, Jingyu Q (2007) Study on use of MSWI fly ash in ceramic tile. J Hazard Mater 141(1):106–114
5. Kockal NU (2015) Optimizing production parameters of ceramic tiles incorporating fly ash using response surface methodology. Ceram Int 41(10):14529–14536
6. Bonczyk M, Rubin J (2022) The application of coal mining waste to the production of construction ceramics: radiological and mechanical aspects. Mater Constr 72(348):e300–e300
7. Almeida EP, Carreiro MEA, Rodrigues AM, Ferreira HS, Santana LNL, Menezes RR, Neves GA (2021) A new eco-friendly mass formulation based on industrial mining residues for the manufacture of ceramic tiles. Ceram Int 47(8):11340–11348
8. Vieira CMF, Monteiro SN (2009) Incorporation of solid wastes in red ceramics: an updated review. Matéria (Rio de Janeiro) 14:881–905
9. Amrani M, Taha Y, El Haloui Y, Benzaazoua M, Hakkou R (2020) Sustainable reuse of coal mine waste: experimental and economic assessments for embankments and pavement layer applications in Morocco. Minerals 10:851
10. Ogunro AS, Apeh FI, Nwannenna OC, Ibhadode O (2018) Recycling of waste glass as aggregate for clay used in ceramic tile production. Am J Eng Res 7:272–278
11. Olgun ASİM, Erdogan Y, Ayhan Y, Zeybek BJCI (2005) Development of ceramic tiles from coal fly ash and tincal ore waste. Ceram Int 31(1):153–158
12. Vasić MV, Mijatović N, Radojević Z (2022) Aplitic granite waste as raw material for the production of outdoor ceramic floor tiles. Materials 15(9):3145
13. Luo Y, Zheng S, Ma S, Liu C, Wang X (2017) Ceramic tiles derived from coal fly ash: preparation and mechanical characterization. Ceram Int 43(15):11953–11966

Chapter 7
Using Mining Waste to Improve Polymer Composite Properties

7.1 Introduction

Polymer matrix composites (PMCs) are fabricated with a variety of fibers, which are bound together to make a polymer matrix. The PMC is fabricated to create a bond between fibres such that the structures made can take mechanical loads. The benefits of PM materials are their lighter weight with high strength. Hence, these materials are applicable in aircraft, automobiles, and for building structures. Due to their corrosion resistance property, these polymer composites are also applicable in marine structures/parts. PMc materials are also durable materials. The materials were adorable as they are made up of waste materials which are plentily available and biodegradable in nature with epoxy as the strengthening agent. Hence, in this chapter, a new class of composite material is explored and their properties are tested and validated as an alternate building material. Also, the sound insulation properties are tested, which will also provide a good solution to the waste management. Table 7.1 shows selection of materials.

As natural fibres are good compared to glass fibres, hence they are considered as alternate sustainable materials. Koizumi et al. [1] have described bamboo fibres as best materials with acoustic property. Putra et al. [2] have explored sugarcane fibre as good acoustic material. Similarly Fouladi et al. [3] recommended coconut fibre, Blunk et al. [4] rice straw and Buratti et al. [5] rice husk.

As, thickness and porosity are considered as the most important property in sound absorption. Hence with all other properties these are also studied.

S. Satapathy et al., *Sustainable Waste Management Practices for the Mining Sector Through Recycling of Mining Waste*, SpringerBriefs in Earth System Sciences,
https://doi.org/10.1007/978-3-031-67932-2_7

Table 7.1 Comparison between natural fiber and E-glass fiber

Parameters	Natural fibers	E-glass fibers
Density	Less	More
Cost	Less	More
Renewability	Yes	Nil
Recyclability	High	Nil
Energy consumption	Less	High
Distribution	Broad	Broad
Abrasion to machines	Nil	Yes
Health risk when inhaled	Nil	Yes
Disposal	Yes	Nil

7.2 Research Methodology

The composites are collected and dried by sun/oven, such that moisture contents can be removed from them. Then these materials well cut to 2–5 mm and prepared mold of both round (65 and height 10 mm) and rectangular shape for impedance tube sound testing and also observing other properties. These molds were made by mixing epoxy resin and hardener in specific ratio (10:1). Silicon sprayer is used before pouring the mixture to avoid sticking to surface. Then these materials were kept by pressing them for 1 day/24 h. Figure 7.1 shows the fabrication process (Fig. 7.2).

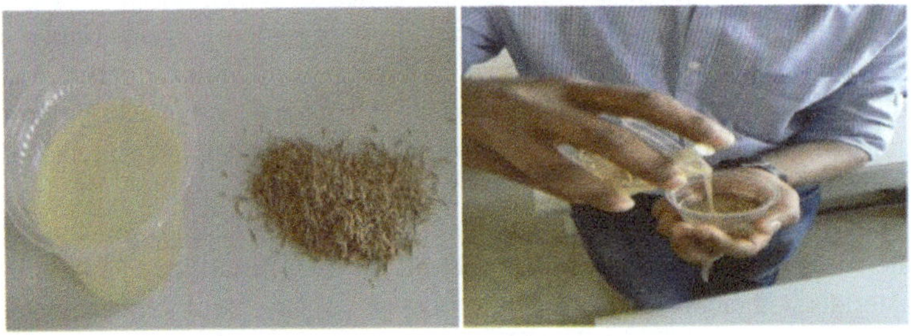

Fig. 7.1 Fabrication process by hand-lay-up technique

Fig. 7.2 Specimen samples after composite fabrication

Then these materials are observed by conducting mechanical, thermal and noise absorbtion test.

7.3 Result and Discussion

The materials are made dog bone shape and tension testing is conducted with the help of Universal testing machine Fig. 7.3. Table 7.2 shows the result of tensile test. Bamboo fibre shows highest tensile stress. Table 7.3 Shows porosity percentage of materials, it is seen that rice husk shows more porosity than other materials and next bambo fibre. Table 7.4 shows thermal conductivity test of the specimen and rice husk having low conductivity compared to others. Figure 7.4 shows a Rockwell hardener tester used to test the harness of 5 samples. Table 7.4 shows the result of hardness testing. Table 7.5 shows Rockwell Hardness's value of all specimens, it was found that bamboo fibre has more hardness compared to others. Table 7.6 shows water absorption test, it was conducted by using below formula

$$WA(\%) = \left(\frac{MW - DM}{DM}\right) \times 100 \tag{7.1}$$

where

MW Moisture weight,

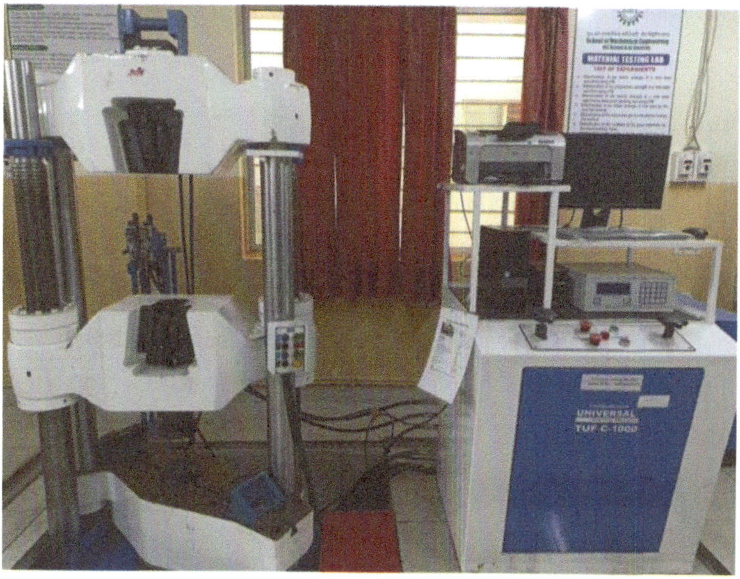

Fig. 7.3 UTM

Table 7.2 Tensile strength and yield strength value of the specimens

Sample No.	Yield strength (MPa)	Ultimate tensile strength (Mpa)
Bamboo fibre	81	168
Coconut fibre	75	162
Sugarcane	32	160
Rice straw	63	154
Rice husk	60	149

Table 7.3 % of porosity of material

Sample No.	Experimental density (g/cm^3)	Theoretical density (g/cm^3)	% of porosity
Rice straw	1.05	1.117	1.76982095
Coconut fibre	1.02	1.066	1.0915197
Sugarcane	0.95	1.043	1.32165868
Bambo fibre	0.94	1.01	7.9306931
Rice husk	0.91	0.97	9.185567

Table 7.4 Thermal conductivity test

No of specimen	Thermal conductivity Keff (W/mK)
Bamboo fibre	0.0913
Coconut fibre	0.086
Sugarcane	0.083
Rice straw	0.075
Rice husk	0.069

DM Dry weight and
WA water absorption.

Sugarcane fibre shows maximum water absorption percentage compared to other composites.

$$\% \text{ of Porosity} = (\text{Theoratical density} - \text{Experimental density}/\text{Theoratical density}) \times 100$$

7.4 Mechanical Property

The noise absorption test was conducted by using an impedance tube and a microphone was inserted and connected with a laptop, one speaker was placed at the one end of the tube (Fig. 7.5). The first test was conducted without any material and

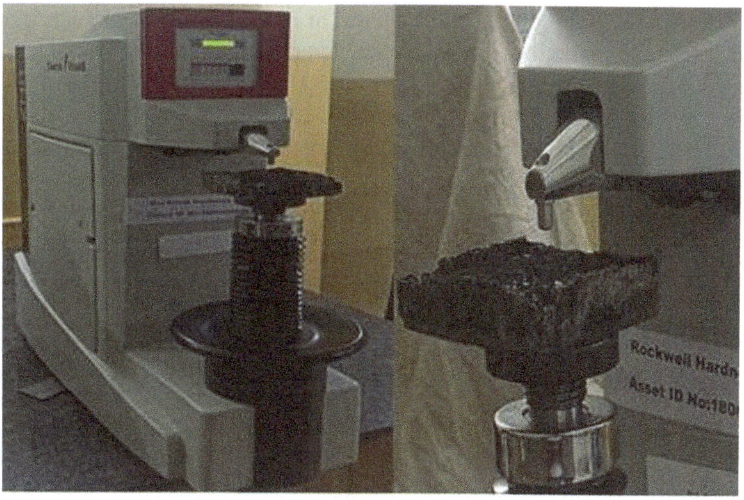

(a) (b)

Fig. 7.4 Hardness test of the specimen

Table 7.5 Rockwell hardness value of all specimens

Load (Kgf)	Type of indenter	Bamboo fibre	Coconut fibre	Sugarcane	Rice straw	Rice husk
100	Jan-16	45.7	43.1	44.6	42.3	42.2
100	Jan-16	45.4	43.3	45.4	43.1	41.2
100	Jan-16	46.7	44.2	45.28	44.1	43.3
100	Jan-16	46.8	44.1	44.5	42.1	39.2
100	Jan-16	47.1	45.3	44.1	42.8	40
Avg. (Hrb)		46.34	44	44.776	42.88	41.175

Table 7.6 Water absorption test

No of specimen	Original weight (gm)	Weight after 48 h in water (gm)	Change in weight (gm)	Water absorb (%)
Bamboo fibre	6.47	6.52	0.05	0.772797527
Coconut fibre	6.48	6.53	0.05	0.771604938
Sugarcane	6.54	6.6	0.06	0.917431193
Rice straw	6.64	6.7	0.06	0.903614458
Rice husk	6.74	6.82	0.08	1.18694362

then all tests were done by placing material at one end of tube in such a manner that outside noise cannot be inferred in the experimental procedure. The procedure continued starting from 500 frequency to 5000 Hz and lab view software was used for analysis, then all the data were recorded for each acoustic material (Table 7.7). From Table 7.7, it was clear that Bambo fibre having noise reduction coefficient more than 0.75, hence it was showing good sound absorbtion property.

Fig. 7.5 Set up for noise test

Table 7.7 Noise absorption test

Sl. No.	Frequency (500–5000) Hz	Sugarcane	Bamboo	Coconut	Rice straw	Husk
1		0.4684008	0.769834	0.542933	0.54452	0.535229
2		0.67865044	0.816539	0.552354	0.691235	0.56437
3		0.38480072	0.812462	0.630177	0.725529	0.639247
4		0.5272028	0.8122096	0.578857	0.720181	0.713655
5		0.50167996	0.847026	0.638618	0.499398	0.620013
6		0.62393352	0.931463	0.648971	0.666814	0.563078
7		0.52403731	0.872328	0.783921	0.659941	0.431867
8		0.43757845	0.822914	0.617584	0.467295	0.735564
9		0.38956495	0.811143	0.663638	0.32545	0.643299
10		0.45992347	0.799658	0.643264	0.595425	0.604477
Noise reduction coefficient		0.499577242	0.82955766	0.6300317	0.5895788	0.6050799

Overall, bamboo fibre shows all good benefits compared to other fibres, but its water absorption percentage is less. Hence, to enhance the composite properties, mining waste was added. Addition of fly ash will enhance mechanical property in fibres, With the addition of red mud thermal stability and flame retard ability property enhanced. Tailing waste improves strength and stiffness property and the addition of slag enhances both mechanical and thermal property. The most importanct thing with addition of mining waste the cost will be less and environment will be safe.

Keskisaari and Kärki [6] have added limestone waste and cartoon cutting waste were reinforced and the composite shows good impact strength and hardness. It was seen that by replacing lime stone waste with IoT the strength of composite increases [7]. Hence an experimentation was conducted and tailing waste was added with bamboo fibre and epoxy and composite fabricated. Same procedure was adopted but fine IoT grains were mixed for proper segregation (Fig. 6a, b).

This composite was further tested for all mechanical, thermal, water absorption and sound absorbtion property. Table 7.8 shows the property of new composite in improved form after adding Iron ore tailing. Table 7.9 shows sound absorption from 500 to 5000 Hz, it shows average noise reduction coefficient was found 0.85174780 $\geqq 0.75$. It shows better results compared to bamboo composites.

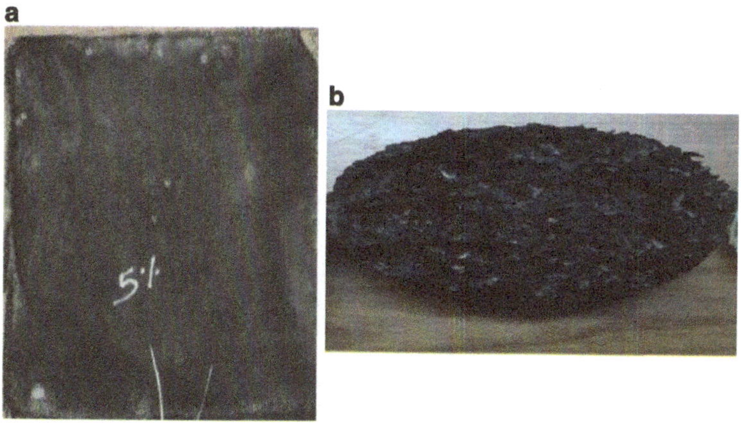

Fig. 7.6 a, b

Table 7.8 Property of new composite

Composite	Yield strength (MPa)	Ultimate tensile strength (Mpa)	Water absorption (%)	Porosity	Avg. hardness	Thermal conductivity Keff (W/mK)
IOT+ bamboo fibre + epoxy	112	134	0.918	9.234	65.23	1.32

Table 7.9 Sound absorbtion property of new composite

Sl. No.	Frequency	New composite
1	500	0.83921061
2	1000	0.75557044
3	1500	0.88172072
4	2000	0.89412285
5	2500	0.97859996
6	3000	0.78085352
7	3500	0.7895731
8	4000	0.79449845
9	4500	0.86648495
10	5000	0.93684347
Noise reduction coefficient		0.851747807

7.5 Conclusion

Composites and smart materials are in high demand due to their application in construction, automobile sector, for packaging of materials etc. When tested before application, certain properties were distracting from the use of composites. Hence, for improving biodegradable composite properties, mining waste will help in improving mechanical properties and usability (applicability) of composites.

References

1. Koizumi T, Tsujiuchi N, Adachi A (2002) The development of sound absorbing materials using natural bamboo fibers. WIT Trans Built Environ 59:157–166
2. Putra A, Abdullah Y, Efendy H, Farid WM, Ayob MR, Py MS (2013) Utilizing sugarcane wasted fibers as a sustainable acoustic absorber. Procedia Eng 53:632–638
3. Hosseini Fouladi M, Nor MJM, Ayub MD, Ghassem M (2012) Enhancement of coir fiber normal incidence sound absorption coefficient. J Comput Acoust20(01):1250003
4. Blunk SL, Yore MW, Summers MD, Lau GK, Tang ST, Jenkins BM (2003) Quality and property changes in rice straw during long term storage. In: 2003 ASAE annual meeting. American Society of Agricultural and Biological Engineers, p 1
5. Buratti C, Belloni E, Lascaro E, Merli F, Ricciardi P (2018) Rice husk panels for building applications: thermal, acoustic and environmental characterization and comparison with other innovative recycled waste materials. Constr Build Mater 171:338–349
6. Keskisaari A, Kärki T (2018) Utilization of industrial wastes from mining and packaging industries in wood-plastic composites. J Polym Environ 26:1504–1510
7. Eugenio TMC, Narciso CRP, Fagundes JF, Henriques AB, Mendes RF (2023) Study on the use of mining waste as raw material for extruded fiber cement production. J Build Eng 63:105547

Index

S. Satapathy et al., *Sustainable Waste Management Practices for the Mining Sector Through Recycling of Mining Waste*, SpringerBriefs in Earth System Sciences,
https://doi.org/10.1007/978-3-031-67932-2

The manufacturer's authorised representative in the EU is Springer Nature Customer Service Centre GmbH, Europaplatz 3, 69115 Heidelberg, Germany. If you have any concerns regarding our products, please contact ProductSafety@springernature.com

Printed and bound by CPI Group (UK) Ltd, Croydon, CR0 4YY
01/10/2026
02225880-0003